T0230773

THE EDITOR

Leon Golberg, M.B., D.Sc., D. Phil., FRCPath., Duke University Medical Center. After receiving a Doctorate in Organic Chemistry from the University of Oxford, Leon Golberg went on to earn a Doctorate of Science in Biochemistry from the University of Witwatersrand, a Master's degree in Anatomy and Physiology from the University of Cambridge, and a degree in Medicine from the University of Cambridge and University College Hospital Medical School in London. Golberg is currently a professor of Community and Occupational Medicine at Duke University Medical Center. Prior to accepting that position, he was responsible for creating the Laboratories of the Chemical Industry Institute of Toxicology and was its first President. In addition, he has been on the faculty of the University of North Carolina and Albany Medical College. A former President of the Society of Toxicology, Golberg established and was the first Director of the British Industrial Biological Research Association. Besides being a member of numerous professional organizations, Golberg has chaired government and other scientific committees. For example, he was chairman of the WHO Scientific Group on Procedures for Investigating Intentional and Unintentional Food Additives; Chairman of the NAS Ad Hoc Subcommittee on Clinical Relevance of Carcinogenicity Testing of Drugs; and Chairman of the Human Effects Subcommittee of the Secretary's (of HEW) Committee on Pesticides. A variety of papers and publications, authored by Golberg, have appeared in the international scientific literature. He serves as Editor, or Associate Editor, of a number of scientific journals in the field of Toxicology.

TABLE OF CONTENTS

Chapter 1

INTRODUCTION

EO presents a major challenge in the area of risk assessment. Here is a versatile and valuable building block of the chemical industry, manufactured in large quantities and finding application in a variety of socially important settings, notably as a sterilant in medical facilities. The different circumstances of production and use of EO have been reflected in different patterns of exposure of the workforce. At one extreme, exposure during manufacture, storage, handling, and transportation has been carefully regulated — especially in recent years — and has been maintained at relatively constant levels. At the other extreme, in situations such as hospital settings, often with outmoded sterilizing equipment and procedures, sharp peaks of exposure of operatives were not uncommon, especially during unloading of sterilizers. In some instances, incidentally exposed employees have constituted the largest group having contact with EO.

EO, the first member of the homologous series of monofunctional aliphatic epoxides is a highly reactive chemical agent possessing a strained ring structure, a property that also finds expression in the biological reactivity of EO. Hence, consideration of the hazards attendant upon the use of EO has to take account of the formation of interaction products with water, chloride ions, and other environmental reactants, as well as the metabolic products of EO formed under biological conditions.

EO reactivity is but one feature of the challenging problem of hazard evaluation of EO. Of fundamental importance is the fact that the administered dose (i.e., level of exposure to EO) is by no means equal to the delivered dose reaching tissues, cells, or macromolecules at sites of critical importance within the body. In addition, difficulties of hazard assessment have been created by emphasis on hemoglobin alkylation as a dose monitor, and by insistence on a radiomimetic action of EO that is said to justify expression of exposure in terms of "rad-equivalents," and of effect in terms of doses of ionizing radiation.

The positive evidence of genotoxicity of EO in lower organisms, and increased SCE in peripheral lymphocytes of workers exposed to EO, will be more amenable to interpretation when the new techniques now becoming available permit measurement of EO-DNA adducts in people. Recent years have witnessed a sharp decline in the levels of occupational exposure to EO, so that measurements of this sort would only be likely to yield positive results in increasingly rare instances of overexposure among sterilizer operators and associated personnel. Haber's Law states that the product of the concentration of a toxicant multiplied by time of exposure is a constant. When one is dealing with reversible (repairable) damage to DNA, the Law does not apply — especially in cases of intermittent exposure, followed by intervals (nights, weekends) when the damage may be partially or wholly restored. In these circumstances, attempts to express intermittent exposure in terms of a calculated, theoretically continuous lifetime exposure are a fallacious basis for mathematical extrapolation from animals to man. Here is but one example where exclusive reliance on statistical manipulations of toxicological data may fail to take into account the underlying biology of the system.

The animal carcinogenicity data illustrate such biological considerations. In view of the high and variable levels of spontaneously occurring mononuclear cell leukemia in Fischer 344 rats, the experimental results lend themselves to different interpretations. In the area of reproductive effects, reports, emanating from Eastern bloc countries and Finland, describe disturbances of menstruation and/or human reproduction, sometimes attributed to extraordinarily low levels of occupational exposure to EO or a variety of other chemicals. Early epidemiological reports of increased incidence of leukemia and gastric cancer in workers exposed to several chemicals, including EO, have not been substantiated in more recent studies.

This confusing situation has been further confounded by efforts to present the issues in oversimplified "black and white" terms. This report attempts to present a critical but balanced appraisal of the available data as a basis for hazard assessment. The conclusions reached reflect a consensus that emerges from the views of independent experts who have attempted to do more than merely recount the available findings.

Since the first draft of this hazard assessment was prepared, the OSHA Hearings have been held on the Notice of Proposed Rulemaking to Revise the Standard for Ethylene Oxide.[2] These Hearings have afforded an opportunity to air many divergent views, and an effort has been made, in this report, to reflect some of the salient issues and points of discussion that have arisen. After consideration of the various viewpoints expressed, no need is seen to modify the original conclusions.

Chapter 2

CHEMICAL AND PHYSICAL PROPERTIES

I. PHYSICAL PROPERTIES

EO is a colorless gas at 25°C with a boiling point of 10.4°C at 1 atm. It is miscible in all proportions with water, alcohol, ether, and most organic solvents. Its vapors are flammable and explosive. Some physical properties of ethylene oxide are summarized in Table 1.

II. STRUCTURE

EO (I) is a strained molecule. Localized molecular orbital studies show that the electron density is strongly polarized toward the oxygen atom.

$$CH_2 \underset{I}{\overset{O}{\diagup \! \! \! \! \diagdown}} CH_2$$

The structural parameters of EO as determined by microwave spectroscopy and electron diffraction are given in Table 2.

III. SPECIFICATIONS OF EO

Commercial EO is essentially chemically pure in terms of purity for most commercial chemicals. Typical specifications are

> Acidity: 0.002 as acetic acid
> Aldehydes: 0.30% as acetaldehyde
> Acetylene: nil
> Water: 0.03%
> Residue: 0.005 g/100 ml
> Color: 10 platinum cobalt
> Odor: nonresidual
> Suspended matter: substantially free

Contaminants of concern for estimating biological or toxicological properties are chlorine-containing C_2 chemicals including vinyl chloride, ethylidene dichloride, chloroethane, and ethylene chlorohydrin. The extent to which these contaminants may be present depends primarily on the particular process used for the manufacture of the chemical. Such impurities, if present at all, are in the range of 1 to 10 ppm concentration and do not significantly affect biological testing results. All modern U.S. manufacturing processes use direct oxidation and halogen-containing impurities are essentially absent. The extent to which ethylene chlorohydrin and ethylene glycol may form under the conditions of a test would have greater potential for modifying the observed biological effect in various test systems than would any contained impurities.

IV. CHEMICAL REACTIVITY OF EO

The principal chemical property of EO which is of concern in estimating biological

Table 1
PHYSICAL CONSTANTS OF EO

Property	Value
Molecular weight	44.05
Boiling point, °C (760 mm Hg)	10.4
Explosive limits in air, volume percent	
Upper	100
Lower	3
Flash point, tag open cup, °C	−18
Freezing point, °C	−112.5
Heat of combustion at 25°C, cal/mol	312.15
Refractive index, n_D	1.3597

Table 2
STRUCTURAL
PARAMETERS OF EO

Bond	Distance, nm
C-C	0.1462
C-H	0.1086
C-O	0.1428
Angle	**Degrees**
HCH	116.9
COC	61.6

risk is the same characteristic which renders the chemical of such enormous value as a chemical intermediate and which is probably also the quality that is responsible for its bactericidal activity, *viz,* its extreme reactivity toward active hydrogen molecules and/or nucleophilic agents. Principally these are amines and alcohols, R-X-H, where X = O or N.

$$RXH + CH_2 \overset{O}{\diagup\!\!-\!\!\diagdown} CH_2 \rightarrow RX\text{–}CH_2\text{–}CH_2OH$$

In the case where R-O-H is water, the reaction leads to hydrolysis with formation of ethylene glycol, and in the special case where HCl is the active hydrogen compound, the reaction leads to the formation of ethylene chlorohydrin.

In all these instances, a quantitative estimation of the rates at which these reactions will take place in biological systems is necessary to assess the "lifetime" of the EO in the biological system, as well as to ascertain the degree to which the observed effects result from the action of a conversion product or metabolite, as opposed to the parent molecule.

Specifically, reaction with primary or secondary amines leads (irreversibly) to the formation of hydroxyethylamines:

$$RNH_2 + CH_2 \overset{O}{\diagup\!\!-\!\!\diagdown} CH_2 \rightarrow RNH\text{–}CH_2\text{–}CH_2OH$$

$$R_2NH + CH_2 \overset{O}{\diagup\!\!-\!\!\diagdown} CH_2 \rightarrow R_2N\text{–}CH_2\text{–}CH_2OH$$

In the case of tertiary amines, quaternization with salt formation occurs:

$$R_3N + CH_2 \underset{\displaystyle \diagup \overset{\displaystyle O}{\diagdown}}{\text{———}} CH_2 \rightarrow [R_3N^+-CH_2-CH_2OH] \; [X^-]$$

The relative rate at which these reactions occur is a function of, inter alia, base strength or nucleophilicity of the amine, the polarity or solvent effects of the medium, and, of course, the concentration of the reactants. (For purposes of estimating the effect on biological systems, the last can usually be ignored since the concentrations of both reaction partners are extremely low in comparison with the high molar concentrations usually employed in chemical reactions for manufacture of other chemicals from EO; in any event, in the simplest form the reactions are true second order, dependent equally on the concentration of EO and the nucleophilic agent.) Aqueous systems, which describe many, if not most biological reaction media, are generally excellent in promoting such reactions since highly polar solvents measurably enhance reaction rates.

The base strength or nucleophilicity of the amine is generally parallel to the equilibrium as determined in the dissociation of the hydrochloride (or other acid salt) of the amine. Aliphatic amines are stronger bases than aromatic amines, and accordingly react more rapidly with EO; these factors do not affect the reversibility of the reactions — they are all essentially irreversible in the usual chemical sense. In biological systems, it can be predicted on the basis of chemical reactivity that aliphatic amine groups (e.g., histidine, glutamine, the NH_2-group of a terminal amino acid, or lysine residues of macromolecules) can, under proper conditions, react with EO irreversibly to form *N*-hydroxyethyl derivatives. Caution must be exercised in drawing conclusions regarding the quantitative aspects of such parallel systems: amine groups in macromolecules are often "masked" because of configurational or "neighboring group" effects in the system. Steric hindrance is a particular form of such modifying effect on relative rates of chemical reactivity by neighboring or adjacent groups, the extent of such blocking being principally dependent on the steric bulk of the chemical group. A substantial degree of alkylation of valine-NH_2 in mouse hemoglobin after exposure to EO has been reported.[21a]

Alcohols (of which water can be a specific case in which R = H) react by a similar mechanism, although as a "base" or "nucleophilic" agent the oxygen atom is far less reactive than the nitrogen atom.

$$R-OH + CH_2 \underset{\displaystyle \diagup \overset{\displaystyle O}{\diagdown}}{\text{———}} CH_2 \rightarrow R-O-CH_2-CH_2OH$$

The parallel derives from the fact that alcohols, particularly aliphatic alcohols, are weak bases, but bases nevertheless by virtue of their available, nonbonding electron pair. In biological systems the relative reactivity of EO with alcoholic OH groups as opposed to aliphatic amine groups of macromolecules is at least an order of magnitude lower. In fact, in the chemical preparation of hydroxyethyl ethers based on aliphatic alcohols, it is necessary to convert the alcoholic OH group to an alkoxide anion, RO^-, by way of a strong base to achieve any significant reactivity at all.

Similarly, mercaptan groups react with EO to form hydroxyethyl thioethers; mercapto (sulfhydryl) groups, as in cysteine, are relatively more reactive than aliphatic hydroxyl groups, but significantly less reactive than the much stronger nucleophilic amine groups.

$$RSH + CH_2 \underset{\displaystyle \diagup \overset{\displaystyle O}{\diagdown}}{\text{———}} CH_2 \rightarrow RS-CH_2-CH_2OH$$

Table 3
RATE CONSTANTS FOR THE HYDROLYSIS
OF EO

Temperature °C	Acidic L/mol/min	Neutral min	Basic L/mol/min
20	0.32	0.000022	0.0034
30	1.00	0.000055	0.01
40	2.5	0.00019	0.0306
131		0.173	

Other active hydrogen groups, such as found in carboxamide ($RCONH_2$) groups, or carboxylic acid (RCOOH) groups are essentially unreactive under normal biological systems, although resourceful chemists have found ways to effect hydroxyethylation of such groups by the use of special systems of high pressure, extreme temperatures, catalysts, conversion of reactive groups to other species, etc.

Chemically, the conversion of EO to ethylene glycol may be accomplished under acidic or basic conditions; in either case the reaction is far more rapid than under conditions of neutral pH, as indicated in Table 3. Similar reactivity is observed in the reaction of EO with hydrogen acids, such as hydrochloric acid, hydrobromic, etc.

$$HCl + CH_2\overset{O}{\overbrace{}}CH_2 \rightarrow HO-CH_2-CH_2Cl$$

An assessment of the magnitude or relative rates of these two reactions must be made in determining the biological implications of testing results of EO, whether such tests are undertaken in animal systems or in vitro systems. Under *most* biological conditions, where the pH may range from 3 to 8.5, the relative rates of these potentially competing side reactions are not highly significant compared to the reaction rate with potent nucleophiles such as aliphatic amine groups, but such competing reactions cannot be ignored.

It is important to note here that, owing to the relatively high solubility of EO in aqueous systems, and owing to the relatively slow change to ethylene glycol or other small-molecule conversion products (the "half-life" is given as 76 hr at 37°C by Osterman-Golkar et al.),[3] the opportunity often occurs for EO to exist in and be transported amidst biological fluids for relatively long periods of time, thus giving opportunity for the molecule to arrive at a reaction site quite distant from the original place of absorption. Estimates of the first order rate constants, assuming that clearance of EO from tissues follows first order kinetics, yield values of 4.6/hr for the mouse[4,5] and 4.2/hr for the rat.[3] The corresponding biological half-lives are 9 and 10 min, respectively. A similar value was deduced for man by Calleman et al.[6] For risk assessment purposes, sound quantitative data are unfortunately not available to permit reliable estimation of persistence data which could lead to an evaluation of the relation between "exposure" doses and "target reactivity" concentrations.

Other chemical reactions of EO which have been well-documented chemically can safely be assumed to be unimportant from a biological standpoint since they occur only under extreme conditions of concentration, temperature, pressure, presence of catalysts, etc., or combinations of these. These include: isomerization, principally to acetaldehyde, polymerization, such as to high molecular weight polyethers, free radical decomposition to methane and carbon monoxide. The fact that such reactions do not occur under conditions even approaching those found in vivo or even in vitro is well documented in the chemical literature.

V. REACTIVITY OF EO WITH BIOLOGICAL NUCLEOPHILES

EO is a direct alkylating agent, reacting with nucleophiles without the need for metabolic transformation. The general mechanisms of reactions of electrophiles with nucleophilic centers in biomacromolecules may involve unimolecular (S_N1) or biomolecular (S_N2) reactions

$$S_N1: \quad R - X \rightarrow R^+ + X$$

$$Y \overset{\sigma^-}{\frown} R^+ \rightarrow R - Y$$

$$S_N2: \quad Y \overset{\sigma^-}{\frown} R \overset{\sigma^+}{-} X \rightarrow R - Y + X$$

where RX is an alkylating agent and YX is a nucleophilic compound. For the overall reaction

$$RX + Y \overset{k_y}{\rightarrow} RY + X$$

k_y depends on the chemical nature of RX and the nucleophilic characteristics of the receptor molecule, the nature of the solvent, and other factors. Reaction rates are expressed by the Swain-Scott equation

$$\log k_y/k_{H_2O} = s \cdot n$$

where s is the substrate constant and n represents the nucleophilic potency of the receptor molecule.

EO acts by the S_N2 mechanism.[7] The value of s for EO has been calculated by Osterman-Golkar et al. as 0.86 at 20 to 30°C.[8] This may be compared with s values of 0.90 for ethyleneimine, 0.93 for epichlorohydrin, and 1.00 for glycidol, all relatively high values. Such alkylating agents with high substrate constants have a greater affinity for centers of high nucleophilicity, such as cysteine sulfhydryl groups in proteins and guanine-N^7 in DNA. This behavior contrasts with the tendency of ethylnitrosourea and isopropyl methanesulfonate (compounds with low s values) to alkylate guanine more efficiently at the all-important O^6 position.[9]

The biological effects of EO correlate with its chemical reaction pattern. The induction of point mutations and chromosomal damage by EO raises the question of nucleophilic selectivity.[10] In a comparison of a series of epoxides, Turchi et al.[11] have confirmed the suggestion of Vogel and Natarajan[12,13] that an increasing ratio of chromosomal aberrations to point mutations parallels an increasing value of s. In fact, the two epoxides with the highest s values were not mutagenic but did induce chromosomal damage. The suggestion is made that different molecular changes in DNA, or damage to chromosomal proteins (through reaction with -SH groups) and/or mitotic apparatus are responsible for point mutations and chromosomal aberrations. It does seem likely that different lesions are involved in these two types of effect.

VI. REACTIVITY OF EO WITH HEMOGLOBIN

The possibility of biological monitoring of persons occupationally exposed to EO was explored by Ehrenberg and his colleagues. The most susceptible amino acids in hemoglobin to attack by EO were found to be cysteine and histidine, and the levels of these alkylated amino acids were measured in mice exposed to [1,2-³H]EO.[4,5] Procedures for quantitating S-methylcysteine and other adducts in hemoglobin have been developed, using GC-MS.[14,15] Support for the idea of using the binding of chemical carcinogens and mutagens to hemoglobin of rats as a dose monitor has also been provided by Truong et al.,[16] Pereira and Chang,[17] Green et al.,[17a] Tannenbaum and Skipper,[17b] Neumann,[17c] and Shugart.[17d]

Implicit in the measurement of alkylated components of hemoglobin is the assumption that the results reflect the extent of alkylation of other macromolecules of interest, such as chromatin proteins and DNA, in various organs and tissues of the body. When [1,2-³H]EO is used, unidentified labeled material becomes adsorbed to DNA.[4] Hence, it is necessary to determine specific alkylation products in order to avoid such contamination. Cumming et al.[18] apparently did not take this precaution when they exposed mice to [³H]EO by inhalation and measured alkylation of DNA in tissues for comparison with alkylation of hemoglobin, at various times after the end of exposure (10 min to 15 days after 1.5 to 3.0 ppm-hr). They reached the conclusion that differences in tissue doses in testis, liver, lung, kidney, and spleen accounted for different degrees of alkylation. Hence molecular dosimetry of chemical mutagens based on the degree of alkylation of remote sites, such as hemoglobin, does not appear to be soundly based. Such doubts are justified especially for germ cells. Whereas red cells are exposed directly to mutagens in blood plasma, germ cells are protected by layers of actively metabolizing cells whose overall impact on the "delivered dose" that ultimately reaches the "target" should not be ignored. Moreover, the extent to which hydroxyethylation of hemoglobin is an indicator for hydroxyethylation of DNA in germ cells is further modified by the fact that DNA in germ cells is extensively bound to protamine, which serves as a sink for EO.[19] Sperm DNA alkylation in the testes appears to be about one fifth that expected on the basis of hydroxyethylated hemoglobin.[20]

The idea was put forward by Ehrenberg and colleagues that red cell histidine alkylation provides a measure of the "integrated" dose or exposure to EO.[3,6,21-24] This concept was based on evidence that the degree of alkylation of amino acids in hemoglobin of circulating erythrocytes was proportional to the amount of cumulative DNA alkylation. The extent of formation of N^3-(2'-hydroxyethyl)histidine in hemoglobin and of N^7-(2'-hydroxyethyl)guanine in livers and testes of male Fischer 344 rats was determined after i.p. injection of [¹⁴C]EO (20.4 μmol/kg) in order to compare directly hemoglobin alkylation with DNA alkylation. The degrees of DNA alkylation were said to be 100% and about 50%, respectively, of the values predicted for liver DNA and testes DNA on the basis of hemoglobin alkylation.[3]

The fact that hemoglobin alkylation is at best a rough approximation of tissue levels of DNA alkylation emerges from studies with mice exposed to methyl bromide[25] or benzyl chloride.[26] As the former authors concluded:

"When hemoglobin alkylation is used for quantitative risk estimations, a correction factor has to be applied by taking into account the difference between the dose in the red blood cells and the dose in the compartments of DNA."

The problem is that the correction factor is different from different compounds, and for the same compound in different species and circumstances.

Quite apart from application for quantitative risk assessment, it is worthwhile to consider the value of measurements of hemoglobin alkylation purely and simply for purposes of biological monitoring of exposure to EO. Thus, detected amounts of alkylated red cell histidine agreed closely with predicted values[6] from male Fischer 344 rats which had been exposed to EO (0, 10, 33, and 100 ppm) for 6-hr/day, 5 days/

week for 2 years.[27] Hemoglobin alkylation has also been used to monitor workers occupationally exposed to EO.[28] On the basis of the reports to date, alkylation of histidine in hemoglobin is a reliable measure of exposure to alkylating agents such as EO. However, alkylation of histidine in hemoglobin is not necessarily indicative of the extent of possible alkylation of cellular macromolecules at more remote sites.

A. Endogenous Formation of EO

Schmiedel and his colleagues[266] found that rat liver microsomes catalyze the formation of EO from ethylene. Rats exposed to ethylene exhale EO;[267] the conversion of the olefin to the oxide is slow, with a distinct saturation pattern.[268,269] Mice exposed to ^{14}C-labeled ethylene or EO developed the same relative amounts of hydroxyethylation products at nucleophilic sites in hemoglobin and DNA.[272] From the degrees of alkylation it was calculated that, at low levels of ethylene, about 8% of the inhaled amount is converted to EO, with a V_{max} for this process corresponding to 4 ppm EO.[272]

There is considerable indirect evidence of the endogenous production of EO, most probably from endogenously formed ethylene,[274] to which there may be a small contribution by ubiquitous environmental sources of the olefin.[273,275] The strongest pointer in this direction is the presence of variable and relatively high background levels of 2-hydroxyethylhistidine in hemoglobin of circulating erythrocytes of people and rats not known to be exposed to EO.[21a,75a,270] A systematic search for 2-hydroxyethylcysteine in the urine, known to be an excretion product in mice exposed to EO,[21] would complement surveys of 2-hydroxyethylhistidine in hemoglobin of normal unexposed people.

Chapter 3

METABOLISM AND PHARMACOKINETICS

As indicated above, EO reacts directly with nucleophiles, without the need for metabolic activation. Since EO is not a substrate for epoxide hydrolase,[29,30] its main pathways of biotransformation would be expected to arise from the reaction of EO with water and chloride ions, resulting in the formation of ethylene glycol and 2-chloroethanol, respectively (Figure 1). The glycol undergoes further metabolism along well-known lines to oxalate, formate, and CO_2.[31-33] The 2-chloroethanol can react directly with glutathione (GSH) or be converted first to chloroacetaldehyde, which then reacts with GSH. The consequence of these changes would be expected to be the formation of S-(2-hydroxyethyl)cysteine, S-carboxymethylcysteine, and their N-acetyl derivatives, all of which would be excreted in urine.

The available evidence bears out some, but not all, of these expectations. Jones and Wells[34] compared the metabolism of 2-bromoethanol and EO, both labeled with ^{14}C, in the rat. They found that these compounds gave rise to similar amounts of S-(2-hydroxyethyl)cysteine and its N-acetyl derivative in urine, possibly because of the partial conversion of bromoethanol to EO. A minor amount of N-acetyl-S-carboxymethylcysteine was present in urine as a metabolite of the proposed intermediate bromoacetaldehyde and of the bromoacetic acid which is known to be formed from the aldehyde. Excretion of radioactivity derived from EO was monitored in expired air for the first 6 hr and in urine for 50 hr. Exhaled EO accounted for 1% and [14]CO_2 for 1.5% of the dose. In urine, the excreted S-(2-hydroxyethyl) cysteine corresponded to 9% of the dose, and its N-acetyl derivative to 33% of the dose of EO.

Turning to the case of 2-chloroethanol formed from EO, oxidation by alcohol dehydrogenase to chloroacetaldehyde[35] is the prelude to conjugation directly with GSH or by way of chloroacetic acid,[36-40] ending up with S-carboxymethylcysteine and its N-acetyl derivative. That formation of S-carboxymethylglutathione in the liver of rats is not the whole story becomes clear from the work of Grunow and Altmann[41] with [1,2-^{14}C] chloroethanol administered orally to rats. Almost the total urinary radioactivity was composed of thiodiacetic and thionyldiacetic acids. At a dose of chloroethanol of 5 mg/kg, the two acids were present in urine in roughly equal amounts; but at 100 mg/kg 70% of the radioactivity in urine was composed of thiodiacetic acid. Strangely, neither unchanged chloroethanol, chloroacetic acid, S-carboxymethylcysteine, its N-acetyl derivative, nor any 2-hydroxyethyl derivative was found in the urine. This report contrasts with the observation by Green and Hathway[42] of N-acetyl-S-(2-hydroxyethyl)cysteine derived from chloroacetaldehyde in rats, and the formation of the same mercapturic acid from vinyl chloride, presumably by way of chloroacetaldehyde, in rats and man.[42-44]

The conversion of 2-chloroethanol into S-carboxymethylcysteine makes it possible to tap into a rich lode of recent research on the metabolism of S-carboxyalkylcysteines in man[45] and, more specifically, of S-carboxymethylcysteine in five mammalian species: rat, rabbit, guinea pig, marmoset, and man.[46] According to Rogers and Barnsley,[45] the ingestion of S-carboxymethylcysteine by man leads to urinary excretion of this compound and its N-acetylated derivative. Such compounds are present in normal human urine, and have been isolated from the urine of patients with cystathionuria. Limitations of space do not permit all the metabolic pathways of S-carboxymethylcysteine to be displayed in Figure 1; as far as human metabolism is concerned, Waring[46] has shown that urinary excretion comprises mainly S-carboxymethylcysteine and its sulfoxide as a major metabolite.

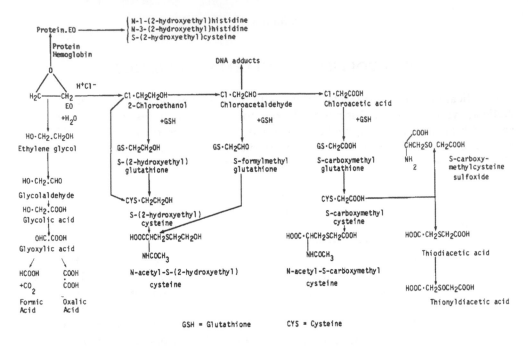

FIGURE 1.   Principal biotransformation products of EO (for references see text).

The initial pharmacokinetic observations on EO were made by Ehrenberg et al.[4] in mice. There are two reasons for suspecting that the mouse may not be representative of other species in its handling of EO. As described above, metabolism of EO (and of vinyl chloride, vinylidene chloride, and 1,2-dichloroethane) leads to the formation of S-carboxymethylcysteine. This product is not deaminated as rapidly in mice as in rats. Hence, in mice given 1,2-dichlorethane i.p., only 33% of the dose is excreted as thiodiacetic acid while 44% of the dose is excreted as S-carboxymethylcysteine.[47] In a preliminary study of species differences in the metabolism of inhaled [1,2-¹⁴C]EO, Tyler[48] used a closed cycle recirculating inhalation metabolism chamber as described by McKelvey and Tyler.[49] He observed metabolic profiles in urine characterized by three acidic and one neutral radiolabeled metabolite fraction. The mouse profile showed two additional fractions, and was thus distinctly different from the rat and (as far as a single determination in each case could ascertain) also from the hamster and guinea pig. Unfortunately the available data do not permit a firm conclusion to be drawn regarding the metabolism of EO in mice as compared to humans.

A further question worthy of discussion is the use of [³H]EO for the study of pharmacokinetics. This radiolabeled compound has considerable drawbacks unless stringent precautions are taken to ensure and establish the purity of the tritiated material, and to ascertain whether tritium exchange with protons takes place in the biological environment. Since such exchange does occur with [³H]EO, purification of specific reaction products is essential before radioactivity is measured. The study of EO pharmacokinetics by Ehrenberg et al.[4] was carried out in mice, using [³H]EO. Some of the assertions and conclusions of this work are dealt with elsewhere in this report. Concentrating here on the pharmacokinetic findings, exposure of male mice (in groups of 1 to 15) to air containing 1.2 to 33 ppm EO for 60 to 107 min, and then analyzing them 0 to 920 min later, revealed rapid pulmonary uptake, tissue distribution and metabolism, and excretion. The biological half-life was calculated as 9 min. Liver, kidney, and lung initially had higher radioactivity than spleen, testis, and brain. Urinary excretion, almost entirely in the first 24 hr, accounted for about 80% of the EO radioactivity believed to be absorbed.

The application of whole-body autoradiography to follow the course of i.v. or inhaled ^{14}C-EO in mice permitted Appelgren et al.[50] to demonstrate high initial levels of radioactivity in liver, kidney, and pancreas. Levels of activity exceeding those in blood were still present at 4 hr in these organs, in the intestine, lung, testis, epididymis, and cerebellum. By 24 hr, radioactivity persisted in liver, intestinal mucosa, epididymis, cerebellum, bronchi, and bone marrow. As might have been expected, the inhaled material gave rise to high levels of activity in the mucosal membranes of the respiratory tract; in other respects the two routes of administration yielded essentially similar results.

Using a closed inhalation chamber system, studies have been carried out in male Sprague-Dawley rats of the pharmacokinetics of ethylene, as a precursor of endogenously formed EO, and of EO as an exhaled reactive metabolite of ethylene.[268,269] The conclusion was that "although ethylene definitely represents a genotoxic hazard" (see however, Reference 273), its low rate of biotransformation to EO, coupled with the distinct saturation pattern of this transformation, precludes the possibility of observing statistically significant tumorigenicity on long-term exposure of rats to ethylene.[271] Continuous exposure of rats for 20 hr to ethylene concentrations > 1000 ppm, which caused maximal V_{max} metabolism to EO, resulted in a plateau of 0.3 ppm EO.[269]

A study of dose-dependent disposition of ^{14}C-EO was carried out by Tyler and McKelvey,[51] using male Fischer 344 rats. Groups of four animals inhaled 10, 100, or 1000 ppm ^{14}C-EO for 6 hr, corresponding to calculated intakes of 2.7, 20.2, and 106.8 mg/kg, respectively. The striking observation concerned the rats exposed to 1000 ppm EO which, during the course of the 18 hr following exposure, excreted proportionately less urinary radioactivity and exhaled more ^{14}CO$_2$ and ^{14}C-EO than did the other two groups. Tissue and carcass levels of radioactivity were correspondingly proportionately higher in the 1000 ppm group. These data suggest that metabolism and excretion of EO involve saturation processes, a finding that has unfortunately not been followed up. Tyler and McKelvey[52] did ascertain that preliminary exposure of rats to 100 ppm EO for 6 hr/day, 5 days/week over 8 to 10 weeks did not materially alter the subsequent disposition of ^{14}C-EO. There was a lower concentration of radioactivity in erythrocytes of pre-exposed rats as compared with controls that had not been subjected to EO inhalation beforehand. In round numbers, the studies by Tyler and McKelvey accounted for about 76% of the radioactivity, excreted as follows: urine, 60%; CO$_2$, 10%; feces, 5%; expired EO, 1%.

The fragmentary character of the available metabolic and pharmacokinetic data in rats and mice is indicated by a study in beagle dogs.[53,54] Here the disposition of intravenously administered EO, ethylene glycol and 2-chloroethanol were compared. The authors found EO to be metabolized mainly to ethylene glycol, so that 7 to 24% of the dose of EO was eliminated as glycol in the urine within 24 hr. Using doses of 25 and 75 mg/kg EO, there was an exponential decline in the plasma levels of EO with a mean rate constant of 0.24 ± 0.008/min and total body clearance of 20.0 ± 5.2 mℓ/kg/min. For ethylene glycol the mean plasma half-life was 221.0 ± 77.7 min and total body clearance 2.13 ± 0.58 mℓ/kg/min. For 2-chloroethanol the elimination half-life and clearance values were 40.8 ± 5.7 min and 10.3 ± 1.7 mℓ/kg/min, respectively.[54]

In the absence of human data on the metabolism and pharmacokinetics of EO, one cannot decide whether the apparently wide divergence between rat, mouse, and dog has important practical implications. Should the dog in fact be more representative of handling of EO by the human body, one would have to reconsider the significance of data for various toxic effects observed in rodents. Although limited data suggest interspecies differences in the metabolic disposition of EO, the rodent (rat) has been selected for quantitative risk assessment because it is the only mammalian system for which chronic data are available. However, as suggested above, dependence on data from a single species can be misleading for purposes of human risk evaluation.

Chapter 4

ACUTE, SUBCHRONIC, AND CHRONIC EFFECTS

I. EFFECTS* OF EXPOSURE TO EO VAPOR: RELEVANCE FOR RISK ASSESSMENT

For humans exposed to EO vapor, sensory warning signs, such as odor, cannot be detected until high concentrations of EO occur. Continued exposure results in olfactory fatigue.[55] Other sensory warning signs, including irritation of the upper respiratory system, have been reported to be undetectable in humans accidentally exposed to high concentrations of EO.[56]

Some of the problems encountered in human exposure to EO have resulted from cutaneous contact. Not only is EO a potent skin irritant,[57] but it has been reported in a study with humans, under laboratory research conditions, to result in delayed hypersensitivity following dermal exposure.[58] However, the authors of this report stated that these types of allergic reactions have not been observed in the workplace with employees who have had frequent contact with EO over a period of many years. Acute allergic responses to EO have been attributed to its use in sterilizing equipment for hemodialysis, peritoneal dialysis, plasmapheresis, and cardiac bypass operations.[58a,h] Reactions among patients undergoing these procedures have included anaphylactoid responses[58b,c] associated with cutaneous reactivity to a conjugate of EO with human serum albumin (EO-HSA), and the presence of IgE and IgG antibody directed against EO-altered HSA, as demonstrated by the radioallergosorbent test (RAST).[58d,e] A recent study[58f] on hemodialysis patients included a group of 7 "reactors" with immediate-type allergic reactions ranging from mild urticaria to severe bronchospasm and hypotension. They were found to have significantly higher levels of EO-HSA bound by serum IgE than in the case of "nonreactor" patients or nonatopic controls. In another study[58g] on dialysis patients, the incidence of positive skin tests to EO-HSA was lower than the proportion of positive RASTs; however, none of the sensitized patients experienced allergic-type reactions during hemodialysis. Marshall et al.[58g] claim that they have demonstrated the immunogenicity of EO, functioning as a hapten, by eliciting high levels of antibody with EO specificity in rabbits immunized with EO-protein carrier conjugates in adjuvant. More recently, Marshall et al.[58i] have reviewed their experience with 25 patients who manifested acute allergic reactions during hemodialysis. IgE antibodies with specificity for EO, and corresponding IgG antibodies, were present in sera from 22 of the 25 individuals who had experienced allergic reactions, but only 5 of 35 unselected patients receiving hemodialysis had such antibodies in their sera. Sera from normal controls or patients allergic to ragweed were entirely negative.

Other reports of human signs from acute exposure to high concentrations of EO have included observations of diarrhea, delayed nausea, and vomiting.[55] A clinical condition compatible with peripheral neuropathy was observed in workers exposed intermittently to high concentrations for up to 2 months (see Chapter 5).[59]

The toxic effects from acute and subchronic EO exposures by several routes of administration for a variety of animal species were summarized by Hine et al.[60] and Glaser.[61] The 4 hr LC_{50} values for rat, mouse, and dog were determined to be 1460, 835, and 960 ppm, respectively.[62]

Results of an 11-week inhalation study in mice at the Bushy Run Research Center (BRRC)[27b] exposed to 250, 100, 50, or 10 ppm of EO, 6 hr/day, 5 days/week, indicated

---

* Exclusive of mutagenic, carcinogenic, or reproductive effects.

a depression in rate of gain in body weight for the 250 ppm group during the last week of exposure. The results of a neuromuscular function test indicated that certain reflex responses and locomotor activities were affected in the highest exposure group and possibly the 100 and 50 ppm groups. There were no accompanying histopathologic alterations in muscle or peripheral nervous tissue. No treatment-related effects were observed for the lower concentration groups. Because of the small sample size, determination of what concentrations were effect or no-effect levels was difficult.

Subchronic BRRC studies and studies performed by Hollingsworth et al.[63] and Jacobson et al.[62] all indicated that only minimal effects occurred in rats exposed to EO at concentrations below 100 ppm. In the BRRC studies, rats were exposed to 450, 150, or 50 ppm for 6 hr/day, 5 days/week, for approximately 8 weeks. Mortality was greatly increased only in the highest exposure group. The predominant cause of death was attributed to vascular damage or nasal cavity obstruction. Other histologic changes noted in this group were lymphoid tissue atrophy and testicular degeneration. These changes were not observed at lower concentrations. Before death occurred in the rats of the 450 ppm group, tremors, convulsions, and paresis of the hindquarters were noted. Only minimal adverse clinical signs, which included paresis of the hindquarters, were noted in the rats of the 150 ppm exposure group. These signs disappeared after the first month of exposure. The only treatment-related effect noted in the 50 ppm exposure group was a transient depression in the rate of gain in body weight.

Hollingsworth et al.[63] and Jacobson et al.[62] reported on the exposure of animals to EO for as long as 6 months by the inhalation route. In rats exposed to approximately 350 ppm, paralysis, and subsequent atrophy of the muscles of the hind limbs were noted. The following manifestations were observed in rats exposed to approximately 200 ppm EO: increased mortality; depressed rate of gain in body weight; organ weight changes; and histologic lesions in the lungs, kidneys, and testes (slight degeneration of the last). Exposure of rats to approximately 100 ppm resulted in a depressed rate of gain in body weight. No adverse effects were observed with exposures to 50 ppm.

The abnormal clinical findings in the subchronic studies at concentrations higher than 100 ppm were not seen in a 2 year inhalation study on rats[27] which were exposed to 100, 33, or 10 ppm of EO for 6 hr/day, 5 days/week. At the 6-, 12-, and 18-month sacrifice intervals, there were no consistent patterns of alteration in urinalysis, hematology, serum clinical chemistry, or organ weight associated with any histologically confirmed organ damage at any exposure concentration. Furthermore, the only histologic lesion seen in the 2-year study, and observed in the subchronic studies at concentrations higher than 100 ppm, was skeletal muscle atrophy. This was present in both sexes of only the 100 ppm exposure group at the 24-month sacrifice interval. No other nonneoplastic, biologically significant effects were discerned from evaluation of the histologic data from the four scheduled sacrifice intervals or from animals that died spontaneously.

II. ANALYSIS OF AVAILABLE DATA

EO has relatively poor sensory warning signs; no odor or irritation of the upper respiratory system can be detected at exposure levels considerably above the current TLV. Abnormal clinical signs in humans were observed in cases where accidental spills or poor ventilation resulted in exposure to very high concentrations of EO. At lower concentrations, clinical signs have not been reported in people exposed for many years or in rats exposed to 10 ppm for most of their lives. Moreover, other than skeletal muscle atrophy, no nonneoplastic, biologically significant, histologic lesions were observed in the 2-year BRRC study on rats exposed to concentrations as high as 100 ppm of EO.[27a] Not enough data are available to assess fully the potential of chronic, low-level EO exposure on neurologic or neuromuscular effects. The NIOSH results of a 2-year study in which the neurophysiology of exposed monkeys was evaluated contributed very little to this end.[122] In conclusion, effects from high-level exposure to EO, either acute or subchronic, are less useful in a hazard assessment in the face of demonstrated mutagenic, carcinogenic, or reproductive effects occurring at lower dose levels.

Chapter 5

NEUROPHARMACOLOGIC AND NEUROTOXIC EFFECTS

I. INTRODUCTION

The first credible clinical description of peripheral polyneuropathy occurring in man as a result of occupational overexposure to EO vapor was not published until 1979.[59] However, it has been appreciated since the 1930s that EO is capable of producing centrally mediated pharmacologic and behavioral effects. At present the number of published observations on exposed animals or clinical cases of overexposure to EO is sufficient to confirm that, under appropriate exposure conditions, inhalation of EO vapor can produce pharmacologic and toxic effects on the nervous system that present themselves as sensorimotor central or peripheral signs and symptoms which may be accompanied by behavioral changes.

The relevant information is assessed concerning the occurrence and significance of neuropharmacologic and neurotoxic effects in both experimental mammals and human populations resulting from exposure to EO. For ease of presentation, the individual studies are tabulated and comparative evaluations with conclusions are presented in the text. In order to evaluate the adequacy of occupational exposure guidelines, particular emphasis has been given to an assessment of potentially adverse neurological effects produced in response to exposure to EO vapor, but routes other than inhalation are considered when they are relevant to the evaluation process.

II. OBSERVATIONS

A. Acute Exposure of Animals

The general effects resulting from exposing animals on single occasions to EO vapor have been well described in several papers (Table 1). The most consistently observed signs in several species include sensory and primary irritation of the eye and respiratory tract, salivation, diarrhea, coma, and death. In review papers, EO is often described as producing central nervous system depression, presumably reflecting the impairment of breathing and the narcosis which are characteristic of acute exposure to high concentrations of EO vapor.[60,64,65] In some instances of death as a result of acute exposure to EO vapor there may be extensive lung injury sufficient to account for death from asphyxia. In other cases, there is minimal lung pathology and respiratory failure may be, at least in part, of central origin.

Signs of intoxication which could result from central actions of inhaled EO are vomiting and convulsions. The appearance of both signs requires exposure to high concentrations of EO vapor. For example, both signs have been described in dogs at EO concentrations of 2830 ppm, with vomiting occurring at 1393 ppm, and neither sign being present at 710 ppm.[62] The suggestion that convulsions may be due to a central effect of inhaled EO is in keeping with the observation that convulsions also occur when EO is given by i.v. injection.[66] More recently, Northup et al.[67] have described ataxia, jerky movements, irritability, and tremor following the i.v. injection of EO into rats, but only at doses close to the LD_{50}.

Convulsions, narcosis, respiratory failure, and vomiting, which may be a result of effects of EO on central nervous system function, occur only as a consequence of exposure to moderately high EO concentrations. Depending on the species, these effects are usually observed when concentrations exceed 200 to 700 ppm.

Table 1

FINDINGS IN ANIMAL STUDIES RELEVANT TO
ASSESSING EFFECTS OF EO ON NERVOUS SYSTEM
FUNCTION BY ACUTE EXPOSURE

Species	Exposure conditions	Signs noted	Ref.
Guinea pig	250—100,000 ppm	Irritation of the eye and respiratory tract; mortality; no effect at 250 ppm	254
Rat	882—2298 ppm for 4 hr	Mortality; respiratory tract irritation; lachrymation; diarrhea; no mention of neurological effects	62
Mouse	533—1365 ppm for 4 hr	Mortality; respiratory tract irritation; lachrymation	
Dog	2830 ppm for 4 hr	Lachrymation; respiratory tract irritation; salivation; vomiting; convulsions	
	1393 ppm for 4 hr	Lachrymation; respiratory tract irritation; salivation; vomiting	
	327 and 710 ppm for 4 hr	No signs noted	
Guinea pig	1320—2850 ppm up to 3 hr	Progressive respiratory distress; death	255
	736—950 ppm up to 3 hr	Respiratory distress; death	
	185—300 ppm for 1 hr	No signs noted	

B. Repeated Exposure of Animals

A sufficient number of animal studies have been conducted to allow a reasonable assessment of the neurotoxic potential of EO by repeated exposure. These are summarized in Table 2, with reference to which the following general comments can be made.

A characteristic neurotoxic effect, predominantly peripheral and affecting the lumbosacral nerves, has been induced in several species by repeatedly exposing them to EO vapor. This takes the form of a paralysis and subsequent atrophy of the muscles of the hind limbs, with an associated decrease in pain perception and reflexes, also in the hind limbs. In those studies where postexposure observations have been sufficiently prolonged, it has been observed that there is a slow but apparently full recovery within 3 to 6 months of the cessation of exposure to EO vapor. The species in which this peripheral neurotoxic effect has been described are rat, mouse, rabbit, monkey, dog, and cat. The guinea pig is either apparently insensitive to the neurotoxic potential of EO, or it occurs only at concentrations where the effects are masked by more serious toxicity.

The atmospheric concentrations of EO which produce or do not result in the induction of paralytic neurotoxicity by repeated exposure are listed in Table 3. These data suggest that the lowest concentrations of EO likely to produce clinical evidence of hind-limb paralysis lie above 200 ppm, with no-effect concentrations having been demonstrated in the range of 100 to 115 ppm. It follows that the threshold concentration for the induction of paralytic neurotoxic effects for most species is in the range 100 to 200 ppm. This accords with the absence of clinical signs of neurotoxicity in Fischer 344 rats exposed to EO vapor at concentrations of 10, 33, and 100 ppm for 6 hr/day, 5 days/week for up to 2 years.[27] Additionally, primates exposed to 50 or 100 ppm of EO

Table 2
FINDINGS IN ANIMAL STUDIES RELEVANT TO ASSESSING EFFECTS ON NERVOUS SYSTEM FUNCTION BY REPEATED EXPOSURE

Species	Exposure conditions	Signs noted	Ref.
Rat, guinea pig, mouse, and monkey	841 ppm: 7 hr each day: up to 8 exposures in 10 days	All died: pulmonary edema, congestion and hemorrhage; hepatorenal injury	63
Rat, mouse, rabbit, and monkey	357 ppm: 7 hr each day: 33 to 54 exposures in 48 to 85 days	High mortality (lung injury found); survivors had paralysis and atrophy of hind-limb muscles; apparently complete recovery 100-132 days after final exposure	
Guinea pig	357 ppm: 7 hr each day: 123 exposures in 176 days	Decreased growth; no neurological effects	
Monkey	357 ppm: 7 hr each day: 38 to 41 exposures in 60 days, or 94 exposures in 140 days	Paralysis with atrophy of hind-limb muscles; poor pain perception in genitalia and hind-limb; decreased knee jerks	
Rat, guinea pig, mouse, rabbit, and monkey	204 ppm: 7 hr each day: 122 to 157 exposures in 176 to 226 days	Decreased growth rate; monkeys had decreased hind-limb reflexes, positive Babinski, and partial paralysis of hind-limbs; rabbits had slight to moderate paralysis; rats died; guinea pigs had no neurological effects; mice apparently had no effects	
Rat, guinea pig, rabbit, and monkey	113 ppm: 7 hr each day: 122 to 157 exposures in 176 to 226 days	Decreased growth rate in rats; no neurological effects noted	
Rat, guinea pig, mouse, and monkey	49 ppm: 7 hr each day: 127 to 131 exposures in 180 to 184 days	No adverse effects noted	
Rat	400 ppm: 6 hr each day: 5 days a week for 6 weeks	Nasal discharge; diarrhea; hind-limb weakness (recovered by 6 months)	62
Mouse	400 ppm: 6 hr each day: 5 days a week for 6 weeks	Weight loss; death	
Dog	290 ppm: 6 hr each day: 5 days a week for 6 weeks	Vomiting; transient hind-limb weakness	
Rat, mouse, and dog	100 ppm: 6 hr each day: 5 days a week for 6 months	Decrease in rate of weight gain; no abnormal clinical signs	
Monkey	50 and 100 ppm: 5 hr each day: 5 days a week for 24 months	No treatment related lesions on microscopic examination of spinal cord, sciatic nerve or ulnar nerve. Axonal dystrophy in distal portion of nucleus gracilis and demyelination in distal portion of fasciculus gracilis	69

Table 2 (continued)
FINDINGS IN ANIMAL STUDIES RELEVANT TO ASSESSING EFFECTS ON NERVOUS SYSTEM FUNCTION BY REPEATED EXPOSURE

Species	Exposure conditions	Signs noted	Ref.
Mouse	236 ppm: 6 hr each day: 5 days each week: up to 10 or 11 weeks	Irwin screen showed abnormalities in tail-reflex, righting reflex, gait, locomotor activity and toe-pinch reflex	27b, 70
	104 ppm: 6 hr each day: 5 days each week: up to 10 or 11 weeks	Abnormalities of righting reflex, gait and locomotor activity	
	48 ppm: 6 hr each day: 5 days each week: up to 10 or 11 weeks	Abnormal gait and locomotor activity	
	10 ppm: 6 hr each day: 5 days each week: up to 10 or 11 weeks	No abnormalities noted	
	0 ppm (unexposed controls)	No abnormalities noted	

Table 3
CONCENTRATIONS OF EO VAPOR PRODUCING AND NOT PRODUCING PARALYTIC NEUROTOXIC EFFECTS IN EXPERIMENTAL MAMMALS

Species	Lowest concentration producing hind-limb paralysis (ppm)	Highest concentration not producing hind-limb paralysis (ppm)
Monkey	204	113
Mouse	357	100
Rabbit	204	113
Rat	357	100
Dog	290	100

for 6 hr/day, 5 days/week for a total of 24 months showed no clinical evidence of neurotoxicity, and peripheral nerve conduction studies demonstrated no differences between the exposed monkeys and the unexposed controls (12 animals per group).[68] Also, no differences were detected on light microscopic examination of the ulnar and sciatic nerves removed from EO-exposed and control monkeys. However, demyelination was seen in the distal portion of the fasciculus gracilis in one of two monkeys of both the high- and low-dose EO groups, and the presence of axonal dystrophy was also noted in the nucleus gracilis.[69] The pathogenesis and significance of these cerebral neuropathological findings, based on an examination of a very small proportion of the exposed primates, is uncertain.

As would be anticipated, when a more detailed clinical examination is made for the presence of early or threshold signs of neurotoxicity, the no-effect level is found to be lower than that observed for obvious paralytic peripheral neurotoxic effects. Thus, in a recent study reported by Snellings,[27b,70] which involved the repeated exposure of mice to EO vapor at various concentrations in the range 10 to 236 ppm, although hind-limb paralysis was not present, there was a dose-related trend in various observations forming part of the Irwin neurobehavioral screen.[71] The threshold for induction of borderline effects was 50 ppm, and the no-effect concentration was 10 ppm (Table 2).

Table 4

CLINICAL REPORTS OF SIGNS AND SYMPTOMS
PRODUCED BY ACUTE OVEREXPOSURE TO EO
VAPOR

Exposure conditions	Observations	Ref.
41 cases of acute EO overexposure resulting from repair work or a major accident	Skin irritation; nausea, recurrent vomiting; diarrhea; headache; vertigo; tachycardia; insomnia	56
Accidental overexposure of a 43-year-old female hospital worker overexposed because of accidental dropping of EO capsule	Nausea; light-headed; brief loss of consciousness; convulsions; chest radiograph normal; inability to perform minor motor tasks for up to 1 week; neurological examination normal at 3 weeks	256

There were no abnormal light-microscopic appearances in the sciatic nerve or gastrocnemius muscle removed from animals at all exposure concentrations.

C. Acute Exposure of Humans

The general signs and symptoms which result from acute overexposure to EO vapor are well known, and have been adequately and extensively documented (Table 4). The most frequently cited effects include eye and respiratory tract irritation, lassitude, nausea, vomiting, diarrhea, vertigo, headache, loss of consciousness, convulsions, and occasionally, disturbances of behavior.

Nausea and vomiting may be particularly troublesome features of acute overexposure to EO vapor. Vomiting is periodic, often with a frequency up to every 30 min, and the nausea and vomiting are often protracted, persisting for up to several days. That nausea and vomiting probably reflect a central neuropharmacologic action of EO is indicated by the observation that they also occur following the percutaneous absorption of EO,[72] and the fact that substantial relief may be obtained from anti-emetics given intravenously.

The dizziness and coma are usually ascribed to a central depressant effect of EO, since they often occur in the absence of significant damage, with adequate aeration of the lungs. However, accompanying these changes in the level of consciousness there may be episodic convulsive movements. That these represent a neuropharmacologic effect of EO is suggested by their occurrence when lung function is not compromised, and the fact that they respond to treatment by intravenously administered anticonvulsants.

A state of excitation and insomnia following acute overexposure to EO had also been described. Recovery from coma, when not associated with marked injury to the respiratory tract, usually occurs within minutes or a few hours. Nausea, vomiting, and convulsions, even if untreated, disappear within a few days.

There is little quantitative documentation concerning the concentrations of EO that cause the above effects. However, it is clear that they can be produced by exposures as short as a few minutes to very high concentrations of the vapor during repair work, or as a consequence of accidents or leaks into confined spaces. Since sensory irritant effects are often present, and because it has been stated that EO could be smelled, the signs and symptoms noted above probably only occur in man at EO concentrations of several hundred ppm.

D. Repeated Exposure of Humans

There is now a sufficient number of observations from various sources to allow a reasonable appraisal of the neurotoxic potential of EO in man as expressed by recurrent exposure to the vapor (Table 5).

Under conditions where there has been inadequate protection of the worker, resulting in the appearance of signs and symptoms of overexposure, the general effects are often similar to those caused by acute overexposure. These include feelings of general ill-health, fatigue, headache, dizziness, nausea, vomiting, and irritation of the eye and respiratory tract. Other findings, which suggest centrally mediated effects of EO by repeated exposure, include nystagmus, ataxia, and slurred speech. The actual concentrations of EO vapor to which individuals have been exposed in order to produce such effects are generally not available, but the work conditions clearly indicate, at least, short-term repeated exposure to several hundred ppm of EO. The lowest reported concentrations producing early symptoms are to be found in a recent study by Garry et al.[73] of 12 individuals in an EO sterilizing facility. They noted incoordination in two, dizziness in three, weakness in four, and difficulty with speech in five; all had irritant effects in addition, and the maximum exposure concentration was stated to be 36 ppm.

The first clear clinical description of a peripheral neuropathy due to repeated exposure to EO was provided by Gross et al.,[59] who examined four workers from a sterilizing facility found to be operating a leaking unit. The cases are reviewed in detail in Table 5. In summary, there was one case of acute encephalopathy with normal nerve conduction studies, two cases having both clinical and neurophysiological evidence of peripheral neuropathy affecting the upper and lower limbs, and one asymptomatic individual who had evidence of sensorimotor polyneuropathy on electrophysiological examination. The amplitude of muscle action potentials, moderate decrease in conduction velocity, and signs of denervation were compatible with an axonal degenerative neuropathy. In the symptomatic cases there was marked subjective improvement within 2 weeks of terminating EO exposure, but over a period of 10 months there was improvement in conduction studies in only one of the three individuals originally found to have abnormalities. The concentrations of EO to which the workers were exposed are unknown but all intermittently smelled the vapor, indicating an exposure on such occasions to at least 700 ppm, the mean detection odor concentration.[62]

A subsequent report, published in 1982,[59a] described the formation of cataracts in three of the four men (Table 5, Cases 2 to 4) seen by Gross et al.,[59] after working on a leaking sterilizer. One of the men (Case 2) developed bilateral cataracts, both anterior vacuolar and posterior subcapsular, over the next $2^1/_2$ years. The two other men both had posterior subcapsular cataracts; in addition, Case 3 had anterior vacuoles in both lenses.

More recently, five further cases of polyneuropathy have been described in young men exposed to EO. The three patients seen by Finelli et al.[59b] were sterilizer operators who regularly smelled fumes at work. Their subacute polyneuropathy was manifested as bilateral footdrop with a slapping gait. Electromyography (EMG) showed denervation potentials. Of the three, two recovered completely in 4 and 7 months, while the third underwent clinical improvement over a 6-month period, but with persistence of denervation potentials on EMG. Kuzuhara et al.[59c] reported the development of sensorimotor polyneuropathy in two sterilizer operators who had been exposed repeatedly to EO for several months, during which time they could smell the gas. Nerve and muscle biopsies were carried out in both patients with, in the second, teased nerve preparation and muscle enzyme histochemistry. The changes seen microscopically and ultrastructurally were consistent with the EMG and nerve conduction findings of axonal degenerative neuropathy. In both patients, symptoms subsided spontaneously within a month and they were able to return to work. Schroder et al.[59d] presented a

Table 5

CLINICAL REPORTS AND OBSERVATIONS ON EXPOSED HUMAN
POPULATIONS REPEATEDLY EXPOSED TO EO VAPOR

Exposure conditions	Observations	Ref.
38-year-old male fumigator of homes and buildings, recurrently overexposed to EO vapor for several days	Weakness; headache; dizziness; nausea; severe vomiting; eye irritation; anxiety; unbalanced and irrational behavior	257
44-year-old male fumigator recurrently overexposed for several weeks	General ill health; nausea; headache; slurred speech; weakness of legs with ataxia; nystagmus; mild disorientation; irrational and aggressive behavior	258
12 employees exposed for 5 months in an EO sterilizing facility: maximum exposure stated to be 36 ppm	Majority experienced irritant effects in eye and upper respiratory tract and nausea; incoordination (2); dizziness (3); weakness (4); difficulty with speech (5).	73
Report of 4 males (age 27-31 years) employed at sterilizing facility with heating EO unit	Case 1. 3 weeks' exposure to leaking unit. Eye irritation; blunting of smell and taste sensation; headache; nausea; vomiting; lethargy; major seizures over a 2-day period; at 2 months normal neurologically and no abnormalities of motor or sensory conduction	59
	Case 2. 3 weeks' exposure to leaking unit. Presented with headache; acral numbness; limb weakness; increasing fatigue. On examination mentally normal; mild weakness of intrinsic muscles of head and foot; heel-shin ataxia; wide-based unsteady gait; muscle stretch reflexes decreased (upper and lower limbs); ankle jerks negative; pinprick and position sensation slightly decreased in fingers and toes; vibration sensation absent in feet. Nerve conduction studies showed general sensorimotor polyneuropathy. Electromyography showed fibrillation of intrinsic foot muscles.	
	Case 3. 2 weeks' exposure to leaking unit. Complained of headache; problems with memory; difficulty in swallowing liquids; cramps; numbness and weakness of extremities. On examination: slurred speech; mild bifacial weakness; moderate weakness of distal muscles; heel-shin ataxia; wide-based gait; muscle stretch reflexes decreased in upper extremities and knees and absent at ankles; vibration sensation decreased in toes. Nerve conduction studies indicated sensorimotor neuropathy. Electromyography showed decreased number and increased amplitude and duration of motor unit potentials.	
	Case 4. Least exposed individual. No symptoms. Neurological examination was normal, but nerve conduction studies indicated a sensorimotor polyneuropathy	
Retrospective morbidity study of 32 male workers occupationally exposed for 5 to 16 years at a stated concentration of 5 to 10 ppm	No significant increase in the incidence of several nervous system diseases in comparison with unexposed controls	77

clinical follow-up study in EO-induced polyneuropathy, with morphometric and ultra-structural findings on a sural nerve biopsy specimen.

In several instances, where there has been a clear indication of repeated exposure to EO vapor, disturbances of behavior have been noted. The descriptions have included a feeling of anxiety, exhibition of unbalanced, aggressive and irrational behavior which is atypical for the individual, and occasionally disorientation. These effects apparently disappear spontaneously over a period of days after the subject is removed from the source of overexposure to EO.

Several epidemiology studies have not mentioned neurologic disease as a complication of occupational exposure to EO.[74-76] In a retrospective morbidity study by Joyner,[77] 37 male workers were reviewed who had between 5 and 16 years occupational exposure to EO at a stated concentration range of 5 to 10 ppm. There was no statistically significant increase in the incidence of several neurological disorders in comparison with nonexposed controls.

E. Conclusions

Based on numerous studies conducted in different laboratories, it may be reasonably concluded that acute exposure of common laboratory mammals to EO vapor produces toxic and pharmacologic signs which can be ascribed to the action of EO on central nervous system function. The principal effects are vomiting, narcosis, convulsions, and respiratory depression. Depending on the species, these signs appear at concentrations of EO vapor greater than 200 to 700 ppm. In the absence of more serious and potentially lethal toxic effects, these signs disappear spontaneously. Similar signs occur in man as a consequence of acute overexposure to EO vapor. Nausea and vertigo may also be present. Little quantitative information exists on the concentrations of EO necessary to induce these signs and symptoms of acute EO intoxication, but the conditions of exposure suggest that, as with animals, the effects only occur at moderately high atmospheric concentrations of EO.

Several studies have shown that repeated exposure of a variety of mammals to EO vapor results in the development of a peripheral neuropathy characterized by paralysis of hind-limb muscles with impairment of sensation and reflexes, and from which recovery occurs. Examination of the exposure data for a variety of species indicates that the concentration of EO vapor that induces paralytic neurotoxic effects lies in the range 100 to 200 ppm. The lowest concentration of EO so far shown to produce abnormal neuromuscular effects is 50 ppm, using a neuromuscular screen in mice; no such effects occur at 10 ppm. A peripheral polyneuropathy has also been described in workers repeatedly exposed to moderately high concentrations of EO vapor, from which symptomatic recovery may occur. Various centrally mediated signs and symptoms have been described in man as a result of repeated exposures; these include nystagmus, slurred speech, nausea, vomiting, anxiety, and irrational behavior. The conditions of exposure suggest that the central and peripheral neurological effects occur as a result of repeated exposure to several hundred ppm of EO, although less severe signs and symptoms may occur at lower concentrations.

Epidemiological studies indicate that neurological complications of occupational exposure to EO are uncommon. A detailed review of the exposure conditions in currently available information on animal studies and exposed humans would indicate that centrally mediated acute and repeated exposure effects, and symptomatic peripheral polyneuropathy, are unlikely to occur if EO exposures are kept below 100 ppm. Taking into consideration the presently available most sensitive indication of early neurological effects, it may be concluded that a significant margin of safety from the neurotoxic effects of EO can be achieved by maintaining EO vapor concentrations at 10 ppm or lower.

In summary, a review of animal studies and clinical reports provides reasonable evidence that EO vapor is capable of causing centrally mediated neuropharmacologic and neurotoxic effects by acute and repeated exposure, and peripheral polyneuropathy by repeated exposure to high concentrations.

Chapter 6

REPRODUCTIVE AND TERATOGENIC EFFECTS

I. INTRODUCTION

This section is concerned with research into the effects of exposure to EO on reproductive performance by males and females, on the process and duration of gestation, lactation, and on the survival, growth, and developmental landmarks of the neonatal animal. Evidence of gynecological and obstetric effects of occupational exposure to EO, such as alterations in menstrual cycles and increased incidence of spontaneous abortions, requires careful examination.

Contributions by other types of studies are important in evaluating any reproductive decrement that may be observed in comparing animals exposed to EO with untreated controls. The issues may be summed up as follows.

1. Does EO reach and penetrate into the reproductive organs? Both autoradiographic[50] and isotope distribution studies[4,18,19] establish localization of EO-associated radioactivity within the testis and epididymis of mice. No direct evidence is available concerning the ovary, but it may be presumed that EO penetrates this organ, too.
2. Does EO cause damage to the gonads? Testicular tubular degeneration and atrophy have been reported in rats and guinea pigs exposed to EO.[63] Toxic effects on male reproductive cells in monkeys exposed to 50 or 100 ppm have been observed.[78,79] The production of dominant lethal mutations in rats and mice[80-82] establishes an effect of EO on sperm. Sega et al.[19] demonstrated maximum alkylation of sperm by ^{3}H-labeled EO in the mid to late spermatid stages. This reflects incorporation of label into protamine rather than DNA, at a period in spermatogenesis that is especially sensitive to EO alkylation.

An excellent account of the reproductive effects of EO in man and animals has been published by Barlow and Sullivan.[83]

II. REPRODUCTION STUDIES

A. Animal Studies

There is only one major published study in which the effects of EO on reproduction were evaluated in experimental animals. This study[84] was conducted on Fischer 344 rats exposed (6 hr/day, 5 days/week) to 10, 33, or 100 ppm EO vapor. There were two air control groups. Both male and female animals were exposed for 12 weeks prior to mating.

During cohabitation the exposure regimen was 6 hr/day, 7 days/week. After copulation was verified by the presence of a vaginal plug, only the females were exposed from day 0 to 19 of gestation (6 hr/day, 7 days/week). The male rats were sacrificed 3 weeks after the mating period. On Day 20 of gestation, exposure was stopped and the dams were allowed to deliver.

At 5 days after parturition, the females (not the pups) were re-exposed to EO 6 hr/day, 7 days/week through Day 21 postpartum. There were no adverse effects on males or females of the F_0 generation. In the group exposed to 100 ppm a larger number of dams did not become pregnant after two mating periods. However, there were no statistically significant differences in the fertility index among groups.

No effects on mortality or body weight gain were observed in the F_0 and F_1 generations. The adverse effects noted in this study were as follows:

1.   Significantly ($p < 0.05$) more dams in the group exposed to 100 ppm of EO had a gestation period longer than 22 days when compared to either air control group.
2.   The median number of pups per litter was significantly lower for the dams in the 100 ppm exposure groups when compared to the control groups. This was considered by the authors to be the major adverse effect found in this study.
3.   The fertility index was lower for the animals in the 100 ppm group when compared to control groups, but the difference was not statistically significant.
4.   The median number of implantation sites per pregnant female was significantly lower for the rats exposed to 100 ppm when compared to the two air control groups.
5.   The median value of the ratio of the number of fetuses born to the number of implantation sites was lower than the value for the dams in either air control group.

These data suggest embryotoxicity or early fetal death at 100 ppm EO.

No significant adverse effects were noted in the dams, or in their progeny, exposed to 33 or 10 ppm EO. Under the conditions studied, 33 ppm can be considered a no-effect dose as it relates to reproduction. Since animals of both sexes were exposed, the mechanism responsible for the adverse effects at 100 ppm EO cannot be ascertained.

B. Human Studies

Yakubova et al.[85] investigated the effect of EO exposure in 282 female workers. The purpose of the study was to evaluate "gynecological disorders as well as the course of pregnancy and childbirth". The workers were classified as "equipment operators" and "laboratory assistants". The control groups were composed of 259 plant administrators in the same factory and 100 other nonexposed workers from other institutions.

The air concentration of EO to which the workers were exposed is not clearly defined, and there is insufficient information concerning the length of exposure of the workers studied. It is implied that the equipment operators were exposed to higher concentrations than the laboratory assistants. It is stated that the level "does not exceed 1 mg/m³" (approximately 0.5 ppm). However, no information is given on whether this was the air concentration at the time of the study, the number of air measurements taken during the study, and whether the air concentration may have been higher prior to initiation of the study. The duration of the study, length of employment, age range of the workers (except in the case of pregnant workers) and parity of the women were not specified.

The major adverse effects reported in this study were

1.   Increased incidence of gynecological disorders ("diseases of the cervix", including erosions; leukoplakia; and "inflammatory disease of the uterus"). The incidence of gynecological disorders was highest among "equipment workers". This is attributed to exposure to higher concentrations of EO for longer periods of time during a working day than in the case of the "laboratory assistants".
2.   Effects on pregnancy such as "threat of miscarriage" and "toxemia" during the second half of the pregnancy were reported to have occurred in "equipment operators" and "laboratory assistants" at a higher frequency than in control groups. No information was provided concerning any correlation between length of employment and incidence and/or severity of these adverse effects. No effects attributable to EO exposure were found in the duration of labor or blood loss during childbirth. Babies born to mothers exposed to EO were not adversely affected.

No conclusions can be drawn from this study due to the numerous shortcomings in study design, the poor presentation of the information, and the vague use of medical terms. Interpretation of the results is rendered more difficult by the fact that, in the factory studied, the EO workers were exposed to high frequency noise and marked changes in temperature. The contribution of these factors to the adverse effects reported is unknown.

Joyner[77] studied the effects of long-term exposure to EO in 37 male workers assigned to EO production units. The purpose of the study was to evaluate overt or subtle toxic effects after long-term exposure of workers to EO. This study was not specifically designed to evaluate effects on reproduction or reproductive organs. The average concentration to which the workers were exposed was 7 ppm and the average duration of exposure was 40 hr/week for 10 $^2/_3$ years. The controls were 41 workers of similar age and duration of employment, not exposed to EO.

There was no statistically significant difference between EO operators and controls in the incidence of "disease of the genitourinary system" (as diagnosed by physicians during an 8-year period.) This was a carefully designed and controlled study. It can thus be concluded that, under the conditions of this study, no adverse effect on the reproductive organs of male workers could be observed.

The epidemiologic investigation of the incidence of spontaneous abortions in Finnish hospital staff exposed to EO[86] is discussed in Chapter 10 and Appendix A.

C. Teratogenicity Studies

The most relevant animal study, involving exposure to EO by inhalation, was conducted by Snellings et al.[87] As judged by maternal survival, litter size, number of implantation and resorption sites, and the extent of preimplantation losses, no adverse consequences and no teratogenic effects were found after exposure of F-344 rats to 10, 33, or 100 ppm of EO vapor for 5 hr/day from day 6 to day 15 of gestation. The only significant adverse effect was a lower body weight of the fetuses from dams exposed to 100 ppm, without embryonic or fetal lethality or abnormalities. The authors concluded that EO was not a teratogen in the species and under the experimental conditions studied.

LaBorde and Kimmel[88] evaluated the teratogenic potential of EO in CD-1 mice. Doses of 75 and 150 mg/kg were administered intravenously at four periods during gestation: Period I (day 4 to 6), Period II (day 6 to 8), Period III (day 8 to 10), and Period IV (day 10 to 12). The dams showed signs of toxicity (tremors, labored respiration, decrease in body weight gain, and some deaths) only at the 150 mg/kg dose level when they were treated during Periods I, III, and IV, but not during Period II. The decrease in body weight gain during gestation in animals which were treated during Periods III and IV was attributable to the smaller number of live pups born to the dams treated at these two time periods.

In the fetuses, the following changes were observed: (1) a lower fetal body weight occurred in pups from dams receiving 150 mg/kg at all four treatment periods; (2) increase in malformed fetuses/litter from dams receiving 150 mg/kg during Periods II and IV. Approximately 19% of pups from dams given 150 mg/kg during Period II had malformations, mainly fusion of the cervical and thoracic arches and fusion and branching of the ribs. Although statistical significance was not reached, there was an increase in malformed fetuses from the dams treated with 150 mg/kg during Period III.

The authors concluded that EO could be considered to be teratogenic when administered intravenously at a dose of 150 mg/kg to mice. At this dose there were marked toxic effects on the dams, including deaths. At a dose of 75 mg/kg the incidence of effects on the dams and the fetuses was similar to that of the control animals. It can thus be stated that, in mice, an i.v. dose of 75 mg/kg could be considered a no-effect level.

Kimmel et al.[89] conducted a similar study in a nonrodent species, the New Zealand White rabbit, to assess the teratogenic potential of EO. The dosages used were 9, 18, and 36 mg/kg on days 6 to 14, and 18 and 36 mg/kg on days 6 to 9 of gestation. The doses chosen were based on a preliminary study in which the maximal tolerated dose was found to be approxiately 40 mg/kg. A "significant trend" towards decreased maternal weight gain was seen during treatment and throughout gestation after treatment on days 6 to 9 or 6 to 14. In this study, no significant effect was seen in the pups from dams receiving any of the doses. An increase in the mean number and percentage of resorptions per litter in dams treated with 36 mg/kg of EO during days 6 to 14 of gestation was the only significant outcome. The authors concluded that i.v. administration of EO to pregnant rabbits does not increase embryotoxicity except at the highest dose level when administered throughout organogenesis and maternal toxicity was observed. Unlike the effects on mice, where a dose of 150 mg/kg was teratogenic, no teratological effects were seen in this study at any dose. An unpublished first draft of the final report of the rabbit study by Jones-Price et al. (dated December 10, 1982) confirms the conclusions presented by Kimmel et al.[89]

D. The NIOSH Study

A useful confirmation and extension of the findings reported by Snellings et al.[84] has been provided in the study by Hackett et al.,[90] hereafter referred to as the NIOSH study. Female rats and rabbits were exposed by inhalation to filtered air or to 150 ppm EO, 7 hr/day, for the following days of gestation (dg): Rabbits — exposure on dg 7 through 19 or 1 through 19; termination dg 30; rats, 3 regimens of exposure: dg 7 through 17, 1 through 16, or 1 through 16 following exposure to EO prior to mating (7 hr/day, 5 days/week for 3 weeks), termination in all cases on dg 21.

By and large, the rabbit experiment revealed no evidence of maternal toxicity, embryotoxicity, or teratogenicity. Rats do not tolerate exposure to 150 ppm EO very well, and maternal toxicity was particularly manifest in the group having 3 weeks of pregestational exposure, followed by 16 days during gestation. Decreased food consumption and reduced gain in body weight was especially prominent in the rats exposed to EO before mating. In litters from this group, the incidence of resorptions was increased and the fetuses were smaller (lower body weight and crown-rump length), as they were — to a lesser degree — in litters from other EO-exposed groups. The observed increased incidence of reduced ossification of skull and sternebrae might have been expected from the general pattern of retarded development in fetuses of EO exposed rats. No overt evidence of teratogenicity was noted. The only inexplicable feature of the results was an increased incidence of hydroureter in litters of rats exposed on dg 7 through 16.

In summary therefore, exposure of rats to 150 ppm EO, under the conditions of the NIOSH study, elicited maternal toxicity and some evidence of embryotoxicity, probably stemming from the adverse effects on the mothers. Thus, the NIOSH study in both rats and rabbits reveals that exposure to EO, even at a level that is so high as to be toxic to pregnant rats, it not teratogenic. These results, taken together with the other evidence from animal studies, strongly suggest that exposure to EO — at levels of 20 ppm or below — is unlikely to exercise any adverse effect on reproduction.

Chapter 7

MUTAGENICITY STUDIES — GENOTOXIC EFFECTS OF ETHYLENE OXIDE IN EXPERIMENTAL MODELS

I. INTRODUCTION

The potential hazard of environmental chemicals with respect to primary damage to DNA may manifest itself as neoplastic disease, birth defects, or various heritable changes.[91] Since any of these effects are highly undesirable consequences of exposure, considerations of DNA damage and its manifestations are an important aspect of the risk assessment process.

The mutagenicity of EO has been extensively reviewed.[92-97] In spite of the literature on the mutagenic properties of EO, existing estimations of actual risk to humans exposed to EO are fraught with uncertainty. This chapter will review the reports on the mutagenicity of EO with the aim of providing a critical evaluation of those studies most applicable to quantitative risk estimation in man.

The data base for submammalian studies of the mutagenicity of EO is thought to be unsuitable for risk assessment in man. Bacterial point mutation data examined alone, for instance, do not make allowance for other important classes of genetic damage readily observed in studies with whole animals. Studies in submammalian systems do not take into account all the information needed to define an effective mutagenic dose in mammals; submammalian systems generally do not possess the active DNA synthesis and repair capabilities of mammals. For these reasons, pertinent studies performed on mammals will be reviewed in depth, while other studies less applicable to risk assessment considerations will be mentioned only briefly.

II. NONMAMMALIAN TEST SYSTEMS

Microbial tests — EO was mutagenic in the Ames *Salmonella typhimurium* test[98,99] and also induced mutations in spores of *Bacillus subtilis*,[100,101] in *Escherichia coli* Sd4[95] and T$_2$h+ bacteriophage.[102]

Neurospora — EO was also mutagenic in the macroconidial strain of *Neurospora crassa*.[103,104]

Plants — EO proved effective in the induction of mutation in barley,[105-107] wheat,[108] rice,[109] and *Tradescantia paludosa*.[110]

Insects — EO induced sex-linked recessive lethal mutations in *Drosophila*.[111,112] Autosomal deletion mutations, lethal mutations, and translocations have also been induced by EO.[113,112]

Cultured eukaryotic cells — A very limited number of reports of the mutagenic effects of EO in cultured mammalian cells are available. A linear dose-related increase in frequency of HGPRT mutants, with no apparent threshold, was observed in Chinese hamster ovary (CHO) cells.[114] The mutation frequency of L51784Y mouse lymphoma cells to resistance to bromodeoxyuridine increased 6- to 14-fold over appropriate controls after growth in polycarbonate flasks sterilized with EO.[115] Unfortunately, no analytical data were presented on the residual concentrations of EO in the media.

III. MAMMALIAN TEST SYSTEMS

A. Unscheduled DNA Synthesis

One method for measurement of DNA damage in the living animals is unscheduled

DNA synthesis (UDS). UDS in germ cells of male mice was measured by Cumming and Michaud,[81] and the most sensitive stages were determined to be mature spermatids and early sperm. Although five daily 8-hr exposures to 300 ppm EO induced UDS, the response observed was not linear with respect to dose, on the basis of the number of exposure days at a given EO level. Strong evidence was presented for a concentration dependent inhibition of DNA repair which may have significantly altered the linearity of the dose-response relationship and the reciprocity of the concentration-time relationship (Haber's law). The concentrations tested had little practical relevance to current workplace exposures.

As an extension of the work of Cumming and Michaud,[81] chemical dosimetry experiments using ^{3}H-labeled EO (by inhalation for 3 ppm/hr) revealed a gradual removal of alkylation products from testicular DNA; 4 days after exposure only 10% of the adducts remained. Thus, there was concordance between the positive UDS response seen in germ cells and removal of adducts from testicular DNA.[20a]

In vitro UDS measurements of effects of exposure to EO in vivo have also been performed. Pero et al.[116] cultured whole peripheral blood, leukocytes, and lymphocytes from factory workers exposed to 0.5 to 1 ppm EO for 0 to 4 years. The UDS induced by treating lymphocyte cultures for 1 hr with 10 μM N-acetoxy-2-acetylaminofluorene (NA-AAF) was determined by measuring incorporation of [3H]-thymidine into DNA. The duration of worker exposure to EO was shown to correlate negatively with the amount of UDS induced by NA-AAF in the lymphocytes. The authors claimed that nontoxic levels of EO had decreased the repair capability of the lymphocytes. Chromosomal aberrations in culture were positively correlated with EO exposure. Further investigations regarding the effects of in vivo exposure of EO showed that the degree of intracellular binding of [3H]-NA-AAF- and NA-AAF-induced UDS varied considerably from culture to culture, but the two parameters were positively correlated. In the EO-exposed group, NA-AAF-induced UDS was significantly reduced when compared to unexposed controls.[117] The differences from unexposed controls were small, however, when compared to variability within the EO-exposed group.

In contrast, in vitro treatment of blood cultures with EO (0.5 to 100 mM for 1 hr) stimulated UDS. At concentrations above 5 mM, EO was cytotoxic and UDS was inhibited. In vitro treatment of lymphocyte and leukocyte cultures with EO also stimulated UDS, but cytotoxicity occurred above 2 mM, and UDS was concomitantly reduced.[116]

Monitoring inhibition of NA-AAF-induced UDS in cultured cells was proposed as a measure of sublethal cellular toxicity. Inhibition of cellular DNA repair capability was considered to be deleterious for exposed individuals.[116] Certainly a careful validation of the methods must be performed before such views can be accepted. As an extension of this work, Hedner et al.[121c] have measured SCEs in human peripheral blood lymphocytes following treatment for 1 hr with NA-AAF or EO, and after a further (18 hr) period of incubation to permit DNA repair to take place. There was a striking difference between the two treatments with respect to DNA repair: with NA-AAF, no significant difference in SCE frequencies, i.e., little or no DNA repair was apparent after 18 hr; with EO, a significant reduction in SCEs had occurred.

B. Cytogenetics: SCES

SCEs have become a popular method of studying primary DNA damage because of the ease and simplicity of scoring exchanges in cultured lymphocytes from animals and man. SCEs can be detected at far lower concentrations of active agents than are usually needed to produce chromosomal aberrations.[118] Nevertheless, the actual mechanism of SCE formation is unknown and the predictive value of SCE as an indicator of genetic damage is still a matter of debate.[119] In one study, short pieces of polyvinyl chloride tubing were sterilized for 90 min with EO. When placed in human skin fibroblast cultures, SCEs were induced. The threshold concentration of liquid EO administered to the cell cultures which was required to cause significant increases in SCE was 36 ppm as determined by gas chromatography; cytotoxicity occurred at 180 ppm. The residual concentrations from the tubing ranged from 12 ppm to 3900 ppm EO, but cytotoxicity was observed only at or above 584 ppm.[120] These findings pointed out the potential hazards of residual EO in sterilized surgical equipment.

In another relevant investigation, 4-month-old New Zealand rabbits were exposed to EO vapor (four rabbits per group) at 0, 10, 50, and 250 ppm, 6 hr/day, 5 days/week for 12 weeks.[121] Blood samples were drawn prior to exposure, at the end of week 1, 7, and 12 of the experiment, and also at 2, 7, and 15 weeks after cessation of exposure. From each group, three rabbits were utilized for serial blood sampling for SCE and hematological assays. A single rabbit was held in reserve, and all four animals were sacrificed after 12 weeks of exposure for glutathione analysis of liver and blood. The negative control rabbits received i.p. injections of balanced salt solution, while the positive control rabbits were treated with 0.5 to 1.0 mg/kg mitomycin C. From each of these rabbits, 50 cells were scored and grown after culturing whole blood for 50 hr. 5-Bromodeoxyuridine ($10^{-5} M$) was added 30 hr prior to scoring of cells; 4 hr prior to scoring, $2 \times 10^{-7} M$ colcemid was added to arrest the cells in metaphase.

After 12 weeks exposure, the 50 ppm and 250 ppm groups exhibited significant increases in the SCE rate per cell (9.47 ± 0.26, $p > 0.05$, and 13.17 ± 0.32, $p > 0.01$, respectively). The 10 ppm group showed a slight increase in SCE per cell (7.85 ± 0.23) compared to the control group (7.26 ± 0.26), but in this instance the control SCE rate per cell was very low compared with that from other sampling times. Overall, 10 ppm was a no-effect level. The elevated SCE rates decreased after termination of the EO exposures, but remained significantly higher than control values even 15 weeks after termination of exposures. Data provided indicate that a cumulative dose of about 660 mg/kg EO is required for induction of significant increases in the baseline SCE rate in rabbits. The calculated exposure rate of rabbits and man at 10 ppm EO was estimated to be 2.2 mg/kg/day for rabbits, and 1.9 mg/kg/day in man. EO had no significant effect on blood glutathione or the hematological indices. Weights of exposed rabbits were reduced, but not in a clear dose-related manner.[121]

In another recent study, SCEs were scored in peripheral lymphocytes from cynomolgus monkeys *(Macaca fascicularis)* which had been exposed to EO for 24 months.[122] Each treatment group comprised 12 adult monkeys at the start of the study. Only two monkeys from each group were sacrificed after 24 months for neuropathological examination; the remaining monkeys were maintained for ongoing quarterly hematology evaluations. Blood samples taken at the end of the 24-month exposure did not reveal abnormalities in red or white blood cell counts. Duplicate blood cultures were incubated with the mitogen phytohemagglutinin for 68 to 74 hr prior to harvesting the cells, which were arrested in metaphase with colcemid; 50 metaphases were scored from each culture. The average numbers of SCEs per metaphase at 50 ppm and at 100 ppm exposure to EO were 10.2 ± 0.5 and 16.8 ± 1.8, respectively, compared to an SCE per metaphase rate of 5.4 ± 0.03 in lymphocytes from unexposed control monkeys.[123] Since no exposures to EO were made below 50 ppm, it is not possible to predict a threshold value for significant increases in SCE frequency in the cynomolgus monkey. Studies on human SCEs are described in Chapter 10.

C. Micronucleus Test

The micronucleus test has been proposed as a simple, effective method for screening for agents which cause chromosomal breakage or nondisjunction of chromosomes. When male CBA mice (three per test group) were exposed to EO i.p. and scored for

Table 1
MOUSE BONE MARROW FINDINGS
AFTER TWO TREATMENTS WITH
SINGLE DOSES OF 0.05 TO 0.20-g EO
PER kg

Total dose (g/kg)	Mean polychromatic erythrocytes with micronuclei	
2 × 0 (control)	0.52	(n = 11)
2 × 0.05	0.31	(n = 7)
2 × 0.10	0.81	(n = 6)*
2 × 0.15	1.38	(n = 4)ᵇ
2 × 0.20	2.48	(n = 8)ᵇ

* $p < 0.05$.
ᵇ $p < 0.001$, n = number of animals.

Table 2
RAT BONE MARROW FINDINGS
AFTER TWO TREATMENTS WITH
SINGLE DOSES OF 0.10 AND 0.15 g EO
PER kg[125]

Total dose (g/kg)	Mean polychromatic erythrocytes with micronuclei	
2 × 0 (control)	0.49	(n = 10)
2 × 0.10	1.08	(n = 4)*
2 × 0.15	0.97	(n = 8)

* $p < 0.05$, n = number of animals.

micronuclei 24 hr later, a significant increase in the number of bone marrow polychromatic micronucleated erythroblasts was observed at 125 mg/kg and above.[124] The authors emphasized the relatively low sensitivity of the micronucleus assay with respect to effective dose.

In another study, two i.v. injections of EO dissolved in water were administered to NMRI mice and Sprague-Dawley rats. From each treated animal, 1000 bone marrow polychromatic erythrocytes were scored for the presence of micronuclei and compared with controls treated with water alone. In both species, doses of 100 mg/kg were sufficient to induce significant numbers of micronucleated erythrocytes. There was a pronounced dose-response relationship, at least with the mice, as shown in Tables 1 and 2.[125]

Bone marrow polychromatic erythrocytes with micronuclei were also scored after inhalation exposure of groups of five male Long-Evans rats to 0, 50, 150, and 1000 ppm EO for 4 hr or 0, 10, 25, and 50 ppm for 4 hr.[98] A positive control group, given triethylene melamine, 0.25 mg/kg i.p., was also scored.

A total of 2000 polychromatic erythrocytes were examined from each rat. In these experiments the threshold for significant increases in the number of micronuclei was 50 ppm. As discussed by Wolman,[93] the dose response relationship in this study was far from linear (Table 2).

D. Cytogenetics: Chromosomal Aberrations

Chromosome damage constitutes a major class of genetic change. Since chromo-

Table 3

MICRONUCLEI IN POLYCHROMATIC ERYTHROCYTES FROM LONG-EVANS RATS EXPOSED FOR 4 HR TO EO[98]

Treatment levels	Micronuclei	SD
High Exposure		
Control	0.12	0.13
50 ppm	0.42[a]	0.22
250 ppm	0.52[a]	0.19
1000 ppm	1.64[a]	0.48
TEM, 0.25 mg/kg	5.50[a]	0.97
Low Exposure		
Control	0.09	0.08
10 ppm	0.12	0.06
25 ppm	0.20	0.07
50 ppm	0.43[a]	0.09
TEM,[b] 0.25 mg/kg	3.41[a]	1.01

[a] $p < 0.05$.
[b] Triethylene melamine.

somal aberrations have been associated with a number of human and animal cancers,[126] it is necessary to evaluate this particular type of endpoint, even though an increased incidence of aberrations does not necessarily predict clinical disease.

In one study, EO dissolved in water was administered orally (9 mg/kg) to 300-g male rats. The animals were killed 24 hr and 48 hr after dosing, and 300 bone marrow cells from 8 rats per group were evaluated for anaphase chromosomal abnormalities. At 48 hr the number of aberrations was considerably less than at 24 hr, but both treatment groups were significantly higher than the control groups.[127] The 48-hr value, however, fell within the range of recorded spontaneous rates. In further work, bone marrow cells from male rats exposed to 3.6 ± 0.6 mg/m^3 (2.0 ppm) and 112 ± 20.2 mg/m^3 (62 ppm) EO for 66 days had chromosomal aberration frequencies of $7.6 \pm 0.1\%$ and $9.4 \pm 0.9\%$ respectively on the 70th day, while only 1.6 ± 0.1 aberrations were observed in untreated controls.[128] Further experimental details were not provided in the translations available. The claim of significant chromosomal effects recorded at 2.0 ppm should be carefully verified in other laboratories.

In experiments conducted by Embree,[98] six male 120- to 150-g Long-Evans rats were exposed to 250 ppm EO 7 hr/day for 3 days. Each rat served as its own control; bone marrow samples were taken before exposure and on the 4th experimental day. There were 20 metaphases photographed and scored for aberrations from each rat bone marrow aspirate. A total of 101/120 aberrations were recorded in the experimental set, while only 7/120 aberrations occurred in the control metaphases, a 14.4-fold increase. Chromatid gaps (30/120 vs. 3/120), chromatid breaks (32/120 vs. 1/120), isochromatid breaks (12/120 vs. 0) rearrangements and exchanges (10/120 vs. 0), dicentrics (5/120 vs. 0) and rings (6/120 vs. 0) comprised the increases in aberration frequency in the metaphases from exposed rats.

NIOSH[123] evaluated chromosomal aberrations after 2 years of inhalation exposures to 50 ppm and 100 ppm EO in the cynomolgus monkey. In their experiments, duplicate 0.5 mℓ whole blood samples were cultured 48 to 52 hr; colcemid was added at a concentration of 0.1 $\mu\ell$/mℓ of culture medium, 4 hr prior to harvest. There were 200 metaphases per culture evaluated for aberrations. EO was evidently toxic, as some of

Table 4

CHROMOSOMAL ABERRATIONS IN
PERIPHERAL LYMPHOCYTES FROM
MONKEYS EXPOSED TO EO FOR 24
MONTHS[122]

Frequency of treatment group	Concentration	Abnormal cells[a]
Control	0	0.75 ± 0.17[b]
EO	50 ppm	2.20 ± 0.38
EO	100 ppm	4.19 ± 1.03

[a] Frequency of cells with one or more chromatid and/or chromosome aberrations per 100 metaphases.
[b] x ± 1 SEM.

the cultures from exposed monkeys did not demonstrate mitogenic activity. The increase in abnormal cells was 3- and 5.5-fold greater than control monkey cells for 50 and 100 ppm, respectively (Table 4). The cells from treated animals demonstrated a dose-related increase in triradials and quadriradials.

Finally, chromosomal aberrations have been induced by EO in a human amniotic cell line in vitro.[129] Exponentially growing cells were exposed for 1 hr to 0 (control), 5, 7.5, and 10 mM of EO which resulted in 100, 58, 25, and 9.2% cell survival, respectively. Chromosome preparations were made 48, 72, or 96 hr after treatment and 50 metaphases were scored for each fixation time (150 metaphases/dose). EO was an efficient inducer of chromosomal aberrations in this cell line and the induction was dose-dependent. The aberrations described were of the chromatid type; chromosome breakage as well as exchanges and complex aberrations were observed.

E. Dominant Lethal Assay

EO has been subjected to the dominant lethal assay, employed as an in vivo measure of mutagenic effects in mammals. Experiments conducted by Strekalova et al.,[128] showed that EO exposure (3.6 ± 0.6 mg/m^3 and 112 ± 20.2 mg/m^3) of male rats for 66 days was associated with 36.4 ± 9 and $47.7 \pm 12.6\%$ embryolethality respectively, when the males were mated to untreated female rats. The control embryolethality was $11.0 \pm 4.5\%$. Postimplantation loss in all groups was practically the same in the high dose, low dose, and control groups: 7.8 ± 2.7, 5.4 ± 1.5, and $6.8 \pm 1.7\%$, respectively. No changes in the function or morphology of spermatozoa could be attributed to EO exposure. The potential significance of data from exposures at 3.6 mg/m^3 (2.0 ppm) cannot be dismissed. However, there is ample reason to suspect the reliability of such information on the basis of other animal data (see below).

Dominant lethal mutations were scored by Cumming and Michaud,[81] who exposed male mice to four 8-hr exposures at 500 ppm, five 8-hr exposures at 400 ppm, and five 8-hr exposures at 300 ppm. The embryolethality was nonlinear with respect to total exposure dose (ppm × time). A peak response of 80 to 90% dominant lethal mutations was observed when four 8-hr exposures of 500 ppm were administered.

EO was also administered by a single intraperitoneal dose of 150 mg/kg to 12-week-old T-stock male mice, and for each male two virgin female (SEC × C57BL)F$_1$ mice were provided. Control males were injected with water. The female mice were replaced when vaginal plugs were observed on each day up to the 22nd day posttreatment. Definite dominant lethal effects were observed in females mated on days 2.5 to 7.5 after treatment, and marginal effects were evident in females mated on days 8.5 to 11.5

posttreatment. These periods correspond to DNA damage which occurred to spermatozoa and late spermatids, with early and midspermatozoa as the most sensitive stages.[82] The dominant lethal effects correlate with studies of DNA alkylation by [³H] EO, in that alkylation occurred most readily in germ cells that were mid to late spermatids at the time of treatments.[19]

In a second experiment by Generoso et al.,[82] hybrid (101 × C3H)F₁ males were given a single i.p. injection of 150 mg/kg EO and mated with one of four strains of female mice 4.5 to 7.5 days after treatment. The range of dominant lethals was 38.1 to 47.0% among the various strains, thus indicating that the ova of these strains varied little in their ability to repair EO-induced lesions in male germ cells. Does the duration of exposure to EO by inhalation affect the incidence of dominant-lethal mutations? This question was addressed by Generoso et al.,[82a] using the same hybrid males as before. They were exposed to 255 ppm EO 6 hr/day, 5 days/week for 2 or 11 weeks and mated (starting on the last day of inhalation) with (C3H × C57BL)F₁ females, the intervals from treatment to mating being 0.5, 1.5, 2.5 and 3.5 days. As compared with 2 weeks of inhalation exposure, 11 weeks produced more dominant lethals (55 vs. 39%) and particularly more dead implants.

A negative response in a dominant lethal assay was reported by Appelgren et al.[50] Single i.v. injections of saline solutions of EO in doses of 25, 50, and 100 mg/kg were administered to groups of five male NMRI mice; provision was made for positive and negative controls. The males were each mated to three female mice for a period of 7 days, and additional females were mated each week for a total of 8 consecutive weeks. A second group of male mice was used for administration of the same doses, but the doses were divided over 2 days. Although the positive controls produced the expected dose response, the experiments with EO yielded results that were not related to dose nor to the stage of spermatogenesis. Appelgren et al.[50] also carried out whole-body autoradiography on mice receiving ¹⁴C-EO by i.v. injection or by inhalation. Although not visible after 2 min, significant radioactivity was detected in the epididymis and testis after 20 min up to 4 hr. It is, of course, likely that at least some of this activity represents biotransformation products of EO.

A group of 15 male Long-Evans rats, 12-weeks-old at the start of the experiment, were exposed to 1000 ppm EO for 4 hr.[98,80] A second group of 10 rats served as positive controls, being treated with a single, i.p. dose of 0.25 mg/kg triethylenemelamine. The 10 male rats which were exposed for 4 hr to filtered room air in an inhalation chamber and subjected to a single i.p. dose of olive oil served as the negative controls. There were two virgin female rats placed with each male rat 24 hr after exposure. At the end of the first week, each male was placed with a new pair of females for a total of 10 consecutive weeks. Evaluation of fetal deaths and total implants occurred on the 17th day after mating. There was a significant increase in the mutagenic index (dead implants per total implants) in the females mated to EO males during the first 3 weeks and 5th week postexposure. There was an increase in total numbers of dead implants in the 2nd, 3rd, and 5th week postexposure. Evidently, EO induced lethal mutations only in sperm stages exposed to EO after meiotic division. No dominant lethal effect was observed in females mated 6 to 10 weeks postexposure. A slight but significant decrease in the fertility index was observed in the 3rd and 4th weeks postexposure, (0.53 in both cases) but this was based upon high control values for the control rats mated during those weeks (0.94 and 1.0, respectively).

F. Heritable Translocation Test

The heritable translocation test has the capability of determining the heritability of chromosomal damage by chemical agents. In a study performed by Oak Ridge National Laboratory (ORNL), groups of 12 T-stock male mice, 10-weeks-old, were in-

Table 5
INDUCTION OF HERITABLE TRANSLOCATIONS WITH EO[82]

Dose, mg/kg/day	No. of progeny tested	No. of partially sterile progeny	No. of sterile progeny	Translocation heterozygotes %
60	406	29	11	9.36
30	456	3	5	1.32
0	822	0	2	0

jected i.p. once per day for 5 weeks with 30 or 60 mg/kg EO. Control males (25) received double-distilled water. The males were mated with (SEC × C57BL)F₁ females during the first week posttreatment. All male progeny were subsequently mated and scored for heritable translocations (Table 5). Significant translocation heterozygotes were induced at both levels.[82]

G. Sperm Motility and Morphology

Cynomolgus monkeys exposed by NIOSH to EO for 24 months and subsequently sacrificed were also examined with regard to testicular weight, sperm motility, and sperm morphology.[78] Monkeys exposed to 100 ppm EO demonstrated significantly decreased testicular weights and epididymal weights; the data for 50 ppm EO are equivocal. Sperm motility was significantly decreased after exposure to 50 or 100 ppm EO. The sperm drive range was also significantly reduced in animals exposed to EO. The incidence of abnormal sperm heads from EO-treated monkeys was twice or more than the observed control levels.

The relationship between the sperm head abnormality data and recognized genetic damage is somewhat uncertain, and such data do not necessarily prove that genetic damage has occurred. Nevertheless, the information indicates definite toxic effects of EO on male reproductive cells.

IV. SPECIFIC-LOCUS TEST

The classical Russell specific-locus test has been applied in order to detect the capacity of inhaled EO to induce gene mutations and multilocus deficiencies at specific genetic sites.[147] After initial trial-and-error at 300 ppm, which killed about one third of the (101 × C3H)F₁ hybrid mice, the level in the first, and then the second and third inhalation studies, was set at 255 ppm for 6 hr/day, 2.9 to 4.4 days/week over a range of 60 to 98 days, accumulating total exposures of 100,800 to 149,940 ppm-hr. Mating with sets of T-stock females continued for about 1 year.

Initially the males showed what appeared to be olfactory disturbances (from EO absorbed on their fur), possibly affecting their well-being or libido. For about the first 3 weeks, the number of females with litters and the average litter size were depressed, but thereafter these indices were normal. Conceptions during the first 7 weeks after the end of exposure yielded only one mutant in 11,525 offspring, while beyond 7 weeks four mutants (two sired by the same male) were found in 59,867 offspring. At the 5% level of significance, the first result rules out an induced frequency that is 0.97 times the spontaneous rate; the second result rules out 6.33, and together they rule out 1.64 times the spontaneous rate.

These findings could be used directly for estimating human genetic risk at exposure concentrations of 0.1, 1.0, 5.0, and 50 ppm TWA₈, assuming 30 years of work exposure to EO, 250 8-hr days/year. The multiples of the spontaneous spermatogonial stem-cell mutation rate ruled out with 95% confidence were 0.08, 0.83, 4.14, and

41.36, respectively. These data for gene mutations induced in stem-cell spermatogonia represent upper limits of human genetic risk from exposure to EO under the specified conditions.

V. RISK ESTIMATION METHODS

A. Alkylation of Proteins and DNA

EO is a highly reactive chemical which readily alkylates DNA and proteins. The amount of alkylation of proteins has been used as a measure of the tissue dose of EO. Male CBA mice were exposed by Ehrenberg and co-workers[4] for 1 to 2 hr with 1 to 35 ppm ethylene-[1, 2³-H] oxide, specific activity 56 mCi/mM, and the rate of alkylation of proteins and DNA was determined. Radioactivity was readily measured in all of the major organs, including lung, liver, kidney, brain, spleen, and testes. Significant amounts of radioactivity were associated with both protein and nucleic acid fractions, and ^{3}H-N-7(2-hydroxyethyl) guanine was detected.

On the basis of the radioactivity measurements made, EO was rapidly distributed roughly equally throughout the body of the mouse including the testes, and quickly detoxified. According to the relationship arrived at, the tissue dose is proportional to the exposure:

$$D = 0.58 \, \mu M \times ppm \times hr$$

No direct mutational endpoints were measured in this study, but estimations of genetic risk due to EO exposure were based upon the proposition that the relationship between the mutation frequency from EO exposure and that induced by X-rays and neutrons is a constant for barley, mice, and even man.

Ehrenberg cites experimental evidence for barley and microorganisms in his arguments that the degree of DNA alkylation is proportional to mutagenic potency which in turn can be expressed in equivalent unit doses of ionizing radiation.[95,130] His proposition assumes that the number of alkylations per unit of DNA associated with an equivalent dose response of γ-radiation (1 rad) is constant in bacteria, plants, and mammals. Ehrenberg calculated the radiological equivalent of EO exposure to be 10 mrad-equivalents per ppm-hr.[130] Assuming a 40-hr week and that an acceptable level of ionizing radiation is 0.1 rad/week, the maximum permissible dose for EO in the workplace would be 2.5 ppm. This estimate may be in error because of the many competing side reactions and the metabolic degradation that occurs in the body during the process of disposition to target molecules in a given tissue. In fact, as discussed in Chapter 11, the entire basis of calculation of exposure in terms of rad-equivalents is probably fallacious.

B. Mutagenesis Data Evaluation

Quantification of genetic risk of EO to man on the basis of animal data requires an evaluation of the existing information to determine mutagenic potency as gauged by various end-points. A summary of relevant published data on the mutagenicity of EO is provided in Table 6.

Experiments with cultured mammalian cells[114] suggest that there is no threshold for the genotoxic action of EO. A threshold of less than 36 ppm for the appearance of SCEs in human fibroblasts is misleading, as the next lowest dose tested was 3.6 ppm.[120] In vitro systems provide information on basic cellular events, but cannot model the complex interactions of the intact organism. Rapid distribution, with a biological half-life of 9 min, active metabolic detoxification, and rapid excretion for EO suggest that an effective threshold for detectable genotoxic effects might be possible. Comparison

Table 6
RELEVANT MUTAGENICITY STUDIES ON EO

Test, species Gene mutations	Dose range	Results (no effect level)	Ref.
Drosophila melanogaster	10—90 mM saline or aqueous solution	Sex-linked recessive lethals, translocations, autosomal deletions (< 10mM)	111, 112, 113
CHO cells, HGPRT locus	0—8 mM	No threshold	114
Rats-dominant lethals	3.6, 112 mg/m³ × 66d (2.0, 62 ppm)	Embryolethality (< 3.6 mg/m³)	128
Mice-dominant lethals	300—500 ppm (8-hr × 5 days)	Embryolethality	81
NMRI mice-dominant lethals	25, 50, 100 mg/kg. i.v.	Equivocal	82
Long-Evans rats-dominant lethals	1000 ppm × 4 hr	Embryolethality	80
Mice strains-heritable translocation	30, 60 mg/kg i.p.	Translocations (< 30 mg/kg)	82
Primary DNA damage			
Mouse germ cell UDS	300—500 ppm (8-hr × 5 days)	No threshold observed	81
Human lymphocytes and leukocytes UDS	0.5—100 mM× 1 hr	No threshold observed	116
Chromosomal effects			
SCEs human skin fibroblasts	3.6—3900 ppm	(< 36 ppm)	12
SCEs rabbit peripheral lymphocytes	10, 50, 250 ppm	(> 10 ppm)	121
SCEs cynomolgus monkey peripheral leukocytes	50, 100 ppm	(< 50 ppm)	122,78
Male CBA mice-micronuclei	50—175 mg/g i.p.	(> 100 mg/kg)	124
NMRI mice and Sprague-Dawley rats-micronuclei	50—200 mg/kg i.v.	(> 50 mg/kg mice < 100 mg/kg rats)	125
Long-Evans rats-micronuclei	10—1000 ppm × 4 hr	(< 50 ppm)	98
Rats-cytogenetics	9 mg/kg oral	Positive	127
Rats-cytogenetics	3.6, 112 mg/m³ (2, 6.2 ppm) × 66 days	(< 3.6 mg/m³)	128
Long-Evans rats-cytogenetics	250 ppm × 7 hr × 3 days	Positive	98
Cynomolgus monkeys-cytogentics	50 ppm, 100 ppm × 24 months	(< 50 ppm)	123

of various routes of exposure, i.e., i.p. or i.v. vs. inhalation, is based upon estimates regarding the minute volume and alveolar absorption. Differences in species, strain and sex further complicate evaluations; reliability of the data reported is also difficult to ascertain.

If SCEs are accepted as the most sensitive genotoxic end-point, the observed threshold for genetic effects in long-term exposure situations is greater than 10 ppm, at least in the rabbit. The threshold for significant induction of micronuclei in the Long-Evans rat was between 25 and 50 ppm for a 4-hr exposure, but definite increases over control values were recorded even at 25 ppm.[98] Recorded data for other endpoints indicate that higher concentrations of EO are required for significant effects (Table 6).

The sensitivity of any particular endpoint for genotoxicity is of considerable importance, but the biological significance of the end-point is of equal importance. SCEs, for instance, occur in the intact unexposed animal at a rather high frequency, but the consequence of such events has not been determined.

Alkylation of DNA has been correlated with a detectable mutation frequency.[130]

Significant alkylation of mouse sperm DNA has occurred when mice were exposed to as little as 1.5 to 3.0 ppm-hr EO.[19] Although the mutation frequency has not been measured at these low exposures, it is assessed to be small but finite. On the basis of alkylation data, the concept of an effective threshold for the genotoxicity of EO is placed in some jeopardy. A paper describing the DNA alkylation by EO at low exposures has recently been submitted for publication.[28] Additional data regarding low levels of exposure would be useful in order to remove the uncertainty that presently exists regarding genotoxic effects at levels approaching those of workplace exposure.

Chapter 8

CARCINOGENICITY STUDIES

I. INTRODUCTION

Historically, low molecular weight, monofunctional epoxides like EO have not been regarded as likely carcinogens.[60] Because they are inherently reactive without metabolic activation, it was expected that such epoxides would react with noncritical nucleophiles before reaching and triggering those critical targets which induce cancer. One critical target, possibly the only one, was assumed to be DNA.

As knowledge concerning the mechanisms of chemical carcinogenesis has grown, the picture has become complicated by the need for further understanding of several aspects, including the following: the potential of any chemical to independently initiate and/or promote carcinogenesis; the kinetics of formation and repair of the putative critical lesion or lesions; the alteration of immune competence, if any, that influences cancer development by the chemical; the capacity of detoxification pathways for endogenous chemicals to affect carcinogenesis up to a potential threshold dosage; and probably other factors as yet unidentified. Knowledge of the species differences between man and test animals with respect to all of these factors, and of differences in exposure patterns, constitutes valuable information for predicting human risk from animal data.

Incidental observations of carcinogenic activity indirectly associated with EO have suggested a need for the evaluation of its carcinogenic potential. Of these incidental observations, three are discussed below. The formal studies to evaluate the carcinogenicity of EO are reviewed, and the weight of the evidence is assessed overall.

II. INDIRECT EVIDENCE FOR EO CARCINOGENICITY

Reyniers et al.[131] described the appearance of tumors in germ-free Swiss-Webster mice accidentally exposed in an isolator to EO-sterilized corncob bedding. This was not the first report of adverse effects of EO-sterilized bedding.[94,132]

In the Reyniers study, 79 adult males and 112 adult females were exposed to the bedding for 150 days. The bedding was changed weekly. A few pups were born during the 150-day period; decreased fertility and increased mortality in males indicated the existence of a problem. The mortality among the males was so high as to preclude further observation. In the 83 remaining females (300 to 900 days of age) indirectly exposed to EO, or to reaction products, a tumor incidence of 86.3% was reported. Multiple tumors per animal were observed; most common were ovarian tumors, malignant lymphomas (possible leukemia), and pulmonary tumors. Reference animals (100 to 600 days of age) had no tumors.

Reyniers et al.[131] pointed out that this was not a systematic study to evaluate EO carcinogenicity. Controls were not matched for age, and the toxic agent was not identified. It is not clear whether free EO was present in this situation; if sufficient off-gassing time occurred between the industrial sterilization of the bedding and its introduction into the isolator, little if any free EO would be expected. Reaction of EO with components of the bedding is likely. The authors reported detection of ethylene glycol in the sterilized bedding, as had Allen et al.,[132] and presumed this would be the reaction product of interest.

Other reviews addressing the potential carcinogenicity of EO either ignore this reference,[60,123,133,134] or minimize its importance except as an indicator of need for a systematic chronic study.[92,96,135] Another type of incidental exposure to EO includes the use of EO-fumigated diet in chronic animal studies.[136] Unfortunately, only one formal comparison is available of cancer incidence in animals fed an EO-treated diet and others fed the same, but untreated, diet. Bär and Griepentrog (1969) saw no increase in tumor incidence in rats fed a diet fumigated weekly with EO in air. In this work, levels of residual EO were measured in the diet to permit an estimation of dose. Similar negative results have been obtained from other long-term feeding studies in which experimental animals received either a diet fumigated with EO, or diet containing known reaction products of EO, e.g., 2-chloroethanol.[38,136,138]

On the other hand, incidental exposure to EO or to some product of EO reaction has been reported to increase the incidence of tumors. Thomas et al.[139] treated mice with three injections of a mammary tumor virus (MTV) which was killed with EO, with the intention of eliciting antibody production and immunity to MTV. Either residual EO or its reaction products (none of which were determined) were credited with causing an increase in tumor incidence. The significance of these positive results is questionable. However, the need for a systematic study of long-term effects of EO is suggested by such evidence.

III. DIRECT (NONINHALATION) STUDIES OF EO CARCINOGENIC POTENTIAL

Walpole[140] exposed 12 rats of unspecified stock to EO in arachis oil by repeated subcutaneous injection. A total dose of 1,000 mg EO/kg body weight was administered over 94 days. There was no reference to other doses in the report, and the dosing schedule was not specified. No local sarcomas were seen at the site of injection when the animals were observed for their lifetimes, the duration of which was not specified. In comparison, other agents did produce local sarcomas at the injection site, e.g., propylene oxide caused sarcomas in 8 to 12 rats receiving a total of 1,500 mg/kg body weight over 325 days (the dosing schedule obviously differing from that of EO). Few experimental details are given which would permit an adequate evaluation of methods or study design. However, the number of animals used was small, and the choice of doses (and/or dosing schedule) appears insufficient to elicit a carcinogenic response in light of a subsequent positive study using this same route of exposure.[142,143] Finally, the route of exposure was inappropriate to permit use of the results for an assessment of potential human risk.

Van Duuren et al.,[141] exposed 30 Swiss-Millerton mice to an estimated dose of 100 mg EO in 10% acetone solution by dermal application 3 times weekly for the entire lifetime. The mean survival time for the EO-treated mice was 493 days. No tumors were reported, and no skin irritation was observed. The number of animals employed, though small, was adequate to detect a sufficient dose of a potent dermal carcinogen. Because of the volatility of EO and its unknown skin penetration, application by painting of an acetone solution on the skin leaves much uncertainty as to the actual EO dose received by the tissue.

Dunkelberg[142] administered EO to female NMRI mice by subcutaneous injection, the route chosen to quantify the applied dose and to make possible the assessment of local tumor development (EO is a tissue irritant), apart from tumors at remote sites. Groups of 100 mice were given weekly injections of 0.1, 0.3, or 1.0 mg EO in 0.1 m*l* tricaprylin. There were 200 untreated controls and 100 tricaprylin-treated controls. The preliminary report[142] presented total mortality and local and remote tumor incidence (each was a composite or cumulative value) in the animals dying before the 92nd week of exposure.

The final report of the study,[143] covers the entire period of 106 weeks, at which time

Table 1

TUMOR INDUCTION BY SUBCUTANEOUS EO INJECTION
IN MICE[a]

Weekly dose (mg)	Mean total dose (mg/animal)	Tumor incidence (%)		
		Sarcomas	Basal cell carcinomas	Malignant lymphomas
1.0	64.4	11		2
0.3	22.7	8	2	2
0.1	7.3	5		2
0[b]	0	4		1
0	0	0		
Historical controls[c]	0	2		

[a] Dunkelberg[143] treated 100 female NMRI mice per group once weekly for 95 weeks and necropsied survivors after 106 weeks.
[b] Controls received the vehicle tricaprylin (0.1 ml/dose).
[c] Historical expectation of tumor incidence from Pott et al.[265]

Table 2

INDUCTION OF TUMORS BY INTRAGASTRIC ADMINISTRATION OF EO IN
FEMALE SPRAGUE-DAWLEY RATS[a]

Dose of EO		No of rats with stomach lesions[b]			
Single dose (2 × weekly) mg/kg	Average total dose mg/kg	Reactive changes[c]	Carcinoma in situ	Fibrosarcoma	Squamous cell carcinoma
7.5	1186	9	4	0	8
30.0	5112	11	4	2	29

[a] Groups of 50 rats were given EO in 1.0 ml salad oil by gavage.[144] No stomach tumors were seen in vehicle or untreated controls.
[b] Rats were classified once only: if squamous cell carcinoma of the forestomach and, say, papilloma occurred simultaneously in the same rat, it was classified under squamous cell carcinoma.
[c] Reactive changes of the squamous epithelium of the stomach comprised hyperkeratosis, hyperplasia, and papillomas. In most cases the lesions listed occurred in the forestomach; however, tumors of the glandular stomach (one a fibrosarcoma) were also seen.

the animals were killed. The results (Table 1) demonstrate a local tumorigenic response associated with EO treatment; the tricaprylin-control tumor incidence slightly exceeds that to be expected from published values. There was no increase in tumors at remote sites.

IV. INTRAGASTRIC ADMINISTRATION OF EO

Dunkelberg[144] elicited tumors and reactive changes in the forestomach of female Sprague-Dawley rats given EO in salad oil by gavage twice weekly for a period of almost 3 years. The results are summarized in Table 2. A possible confounding factor was an outbreak of pneumonia during weeks 79 to 82 of the experiment, as a result of which the animals received 2 × 100 mg chloramphenicol s.c. and tylosin tartrate in drinking water for 3 weeks, during which time administration of EO was interrupted. For the most part the pathological effects of EO were manifested in the forestomach. In 15 rats receiving 30 mg doses of EO, the resulting tumors invaded adjacent organs

Table 3A

INCIDENCE OF MONONUCLEAR CELL LEUKEMIA IN
FISCHER 344 RATS EXPOSED TO EO: BRRC
INHALATION STUDY[a]

	Tumor incidence (and %) at EO concentration (ppm)				
Sex	0	0	10	33	100
Male	5/48 (10)	8/49 (16)	9/51 (18)	12/39 (31)	8/30 (30)
Female	5/60 (8)	6/55 (11)	11/54 (20)	14/48 (29)	15/26 (58)[b]

[a] Groups of 120 rats per sex per group were exposed to EO by inhalation for 6
    hr/day, 5 days/week, for 24 months (females) and up to 26 months (males).
    Tumor incidence data are for those animals killed after these times.[27,27a]

[b] Statistically significant ($p < 0.0001$).

and metastases were frequently present. An unspecified number of tumors occurred in
the glandular stomach.

V. FIRST INHALATION CARCINOGENICITY STUDY

A chronic EO inhalation study [referred to as the Bushy Run Research Center
(BRRC) study] was carried out by Snellings et al.[27,27a] The study has been reviewed by
NIOSH[123] and OSHA.[97] The study met or exceeded the requirements of Good Labo-
ratory Practices.[145]

Groups of 120 male and 120 female Fischer 344 rats were exposed to 0, 10, 33, or
100 ppm EO for 6 hr/day, 5 days/week for a total of 2 years. Because of a virus
infection (see below), exposures were discontinued for a 2-week period in the 15th
month of the study. There were two control groups exposed to room air. Interim kills
were done at 6, 12, and 18 months. Histopathology was performed at all dose levels in
the animals killed at the end of the study, in the controls and top dose (100 ppm)
groups at interim kills, and in all animals found dead or killed moribund.

Weight gain was depressed in females of the 100 ppm exposure group as early as 3
weeks and became most striking after 20 months. A nonsignificant depression of
weight gain occurred in the 33 ppm group. EO exposure led to increased mortality,
depression of red cell counts and hemoglobin levels in females at 100 ppm EO, and
mild skeletal muscle atrophy, thought to be nonneurogenic in origin, at 24 months.

An outbreak of SDA (sialodacryodenitis) in all groups of rats was associated with
an increased mortality beginning in the 15th month of the study; mortality was elevated
in the 33 ppm groups, and was greatest in the 100 ppm groups. Females were more
sensitive than males. After cessation of EO exposures for 2 weeks, mortality declined
after the 17th month, the clinical appearance of the animals returned to normal, and
body weights began to increase. All surviving females were killed at 24 months, and
males at 26 months, after the start of the study.

After 24 months of exposure, female rats in the 100 ppm group had a greater incidence of mononuclear cell leukemia (MCL) than females in either control group (Table
3A). Among females exposed to 100 ppm EO, the cumulative percentage developing
MCL was significantly greater than in control groups. At 33 ppm the cumulative incidence of female MCL was greater than one control group or the combined control
groups, but the difference with respect to the second control group did not reach statistical significance (Table 3B). The positive, linear dose-response of MCL incidence in
females exposed for 24 months was significantly correlated with exposure ($p < 0.01$)
with a correlation coefficient r > 0.99. Mortality-adjusted trend analysis revealed either

Table 3B
CUMULATIVE PERCENTAGE OF FISCHER 344 RATS DEVELOPING MONONUCLEAR CELL LEUKEMIA DURING CHRONIC EO EXPOSURE: BRRC INHALATION STUDY[a]

	Cumulative percent of animals developing tumors at exposure concentrations (ppm)				
Sex	0	0	10	33	100
Male	36	40	41	55	61
Female	21	28	36	50[b]	79[c]

[a] Groups of 120 rats per sex per group were exposed to EO by inhalation for 6 hr/day, 5 days/week, for up to 26 months. Cumulative percentages of animals developing tumors include all animals killed at interim times or study termination, as well as those animals found dead or killed moribund during the course of the study. Data are computed from life-table analysis.[27,27a]

[b] Statistically significant compared only to the first control group or to combined controls ($0.05 > p > 0.01$).

[c] Statistically significant compared to either control group or to combined controls ($p < 0.0001$).

Table 4
INCIDENCE OF PERITONEAL MESOTHELIOMAS IN MALE FISCHER 344 RATS: BRRC INHALATION STUDY[a]

	EO concentration (ppm)				
Incidence parameter	0	0	10	33	100
Incidence (and %) in rats exposed for up to 26 months[b]	1/48 (2)	1/49 (2)	2/51 (4)	4/39 (10)	4/30 (13)
Cumulative percentage of rats developing tumors[c]	5	5	8	22	47[d]

[a] Groups of 120 rats per sex per group were exposed to EO by inhalation for 6 hr/day, 5 days/week for up to 26 months.[27,27a]

[b] Data are for animals killed at termination of the study.

[c] Data are computed from life-table analysis and include all animals killed at interim times or study termination, as well as those animals found dead or killed moribund during the course of the study.

[d] Statistically significant compared to either group or to combined controls ($p < 0.001$).

an increased rate or an increased incidence of MCL in male rats ($p < 0.01$) and an increased incidence of MCL in females ($p < 0.0001$).

In male rats surviving to the termination of the study, the incidence of peritoneal mesothelioma was elevated in the 33 and the 100 ppm exposure groups (Table 4). By life table analysis, the cumulative incidence of mesothelioma in males exposed to 100 ppm was statistically greater than controls (Table 5). Mortality-adjusted trend analysis indicated a significant relationship ($p < 0.0001$) between EO exposure and development of peritoneal mesothelioma.

Exposure to EO increased the rate of development, but not the total incidence, of pituitary adenomas in male and female rats. Trend analysis showed significant trends for males ($p < 0.01$) and females ($p < 0.0001$) for the early development of this tumor type at 100 ppm.

Table 5
INCIDENCE OF NOTEWORTHY LESIONS IN MALE
FISCHER 344 RATS: NIOSH STUDY[a]

Lesion	Incidence (and %)[b] EO (ppm):		
	0	50	100
Mononuclear cell leukemia[b]	24/77 (31)	38/79 (48)[c]	30/76 (39)
Peritoneal mesothelioma	3/78 (4)	9/79 (11)	21/79 (27)[d]
Brain			
Gliosis	0/76 (0)	2/77 (3)	4/79 (5)
Glioma (mixed cell)	0/76 (0)	2/77 (3)	5/79 (6)[c]
Astrocytoma	0/76 (0)	0/77	0/79 (0)
Pituitary			
Carcinoma	4/73 (5)	0/66 (0)	0/67 (0)
Adenoma	44/73 (60)	20/66 (30)[c]	21/67 (31)[c]

[a] Groups of animals exposed by Lynch et al.[122,122a] for 24 months, 7 hr/day, 5 days/week.

[b] Diagnosis based on histological evaluation of spleens.

[c] Statistically significant difference from controls, $p < 0.05$.

[d] $p < 0.01$.

Examination of brain tissue at all treatment levels revealed the presence of primary brain neoplasms.[146,146a] The total yield of primary brain neoplasms reported in the BRRC study is indicated in Tables 6 and 7. Statistical analyses of the data have been carried out by Sielken,[222] (Appendix C) who concluded that the 10 ppm results were indistinguishable from those of the control groups, but that the probability of a response was appreciably increased at the 33 ppm level.

The number of female rats with multiple malignant neoplasms was statistically greater than controls at 33 or 100 ppm EO, and the incidence of multiple neoplasms (benign and malignant combined) in females was statistically greater than in controls at all three exposure concentrations (10, 33, and 100 ppm). This was not the case in male rats.

VI. SECOND INHALATION STUDY

The results of a study sponsored by the National Insitute for Occupational Safety and Health (NIOSH) have been published by Lynch et al.[122,122a] In this NIOSH chronic inhalation study, groups of 80 male Fischer 344 rats, and groups of 12 male cynomolgus monkeys *(Macaca fascicularis)* were exposed to EO at 0, 50, and 100 ppm for 7 hr/day, 5 days/week for 24 months. Rats were killed at 24 months. Cytogenetic assessments (both chromosomal aberrations and sister chromatid exchanges) and sperm evaluations (motility, density, and morphology) were carried out on all monkeys surviving 24 months, at which time two monkeys from each group were killed for detailed necropsies (including neuropathology).

During the study, rat exposures were discontinued for 2 weeks due to an outbreak of *Mycoplasma pulmonis* which occurred after 16 months and which was treated with tetracycline. Hematologic evaluation revealed a significant elevation in lymphocytes in EO-treated rats, while total white cell counts were unaffected. In monkeys, white blood cell and lymphocyte counts were variable and did not follow the pattern in rats. The incidence of mononuclear cell leukemia (MCL) at the 24-month kill was positively correlated with exposure concentration (0, 50, 100 ppm: 33, 42, 67%, respectively). When cumulative incidence was examined (Table 5), a statistically significant relation-

Table 6
FREQUENCY OF PRIMARY BRAIN NEOPLASMS IN
F344 RATS: BRRC INHALATION STUDY[a]

Sex	Exposure level (ppm)				
	100	33	10	0(CI)	0 (CII)
18 Month Sacrifice					
Male	0/20[b]	1/20	0/20	0/20	0/20
Female	1/20	0/20	0/20	1/20	0/20
24 Month Sacrifice					
Male	3/30	1/39	0/51	1/48	0/49
Female	2/26	2/48	0/51	0/60	0/56
Dead/Euthanatized Moribund					
Male	4/49	3/39	1/28	0/30	0/29
Female	1/53	1/31	1/24	0/18	0/20
18 Month, 24 Month Sacrifices and Dead/Euthanatized Moribund (Combined from Above)					
Male	7/99	5/98	1/99	1/98	0/98
Female	4/99	3/99	1/95	1/98	0/96
2-Year Study (Combined 6, 12, 18, 24 Month and Dead/Euthanatized Moribund Animals)					
Male	7/119[c]	5/118	1/119	1/118	0/118
Female	4/119	3/119	1/115	1/118	0/116

[a] Groups of 120 rats per sex per group were exposed to EO by inhalation for 6 hr/day, 5 days/week for up to 24 months (females) and 26 months (males).[27,27a]

[b] Numerator equals the number of brains with primary neoplasms. Denominator equals total number of brains examined microscopically.

[c] See footnote b, Table 7.

ship to exposure was apparent at 50 ppm but not 100 ppm EO. Peritoneal mesothelioma was significantly increased at 100 ppm. There was no evidence of leukemia in the monkeys.

Tumors of the central nervous system, made up entirely of mixed cell gliomas, were increased in an exposure-related manner. The cumulative incidences were 0/76 (0%), 2/77 (2.6%), and 5/79 (6.3%) in the control, 50 and 100 ppm exposure groups, respectively. Interestingly, no astrocytomas were reported, while pituitary tumors — especially adenomas — were reduced considerably.

VII. THIRD INHALATION STUDY

A well-conducted subchronic inhalation study has been carried out by BRRC in B6C3F$_1$ mice.[27b] Groups of B6C3F$_1$ mice were exposed to 236, 104, 48, 10, and 0 ppm EO for 6 hr/day, 5 days/week for a total duration of 10 weeks (males) or 11 weeks (females). Exposure at the highest dose revealed some slight hematological effects and organ weight changes. Reference is made to unpublished work in DBA mice exposed

Table 7

FREQUENCY OF PRIMARY BRAIN NEOPLASMS TYPES IN F344 RATS: BRRC INHALATION STUDY[a]

Neoplasm type	Exposure level (ppm)				
	100	33	10	CI	CII
Males					
Granular cell tumor	1/119[b]	1/118	1/119	0/118	0/118
Astrocytoma/oligodendroglioma/mixed glioma	5/119	2/118	0/119	1/118	0/118
	6/119[c]	3/118[c]			
Malignant reticulosis-microglioma	1/119	2/118	0/119	0/118	0/118
	0/119[c]	1/118[c]			
Females					
Granular cell tumor	1/119	1/119	0/118	1/118	0/116
Astrocytoma/oligodendroglioma/mixed glioma	2/119	2/119	1/118	0/118	0/116
Malignant reticulosis-microglioma	1/119	0/119	0/118	0/118	0/116

Note: (Combined data for 6, 12, 18, 24 months and dead/euthanatized moribund animals).

[a]   Groups of 120 rats per sex per group were exposed to EO by inhalation for 6 hr/day, 5 days/week for up to 26 months.[27,27a]

[b]   Numerator equals the number of brains with neoplasm. Denominator equals total number brains examined microscopically. Although 6- and 12-month sacrificed animals are included, no brain neoplasms were discovered in these groups. One can eliminate the 6- and 12-month animals by subtracting 20 from each denominator.

[c]   Frequency reported by Garman et al.[146a]

to 200 ppm EO for 6 months, in which neither these effects nor histopathological evidence of irritative effects on the respiratory passageways was seen. In the study with B6C3F$_1$ mice (see Table 1), particular attention was paid to reflex responses and other possible neuromuscular effects. There was some evidence for a dose-related trend in response to tests of impaired locomotor function, but the observations were unsupported by any histopathological changes that could be discerned in muscle or central or peripheral nervous tissue.

A chronic inhalation study has been in progress at Batelle, Northwest in B6C3F$_1$ mice, sponsored by the National Toxicology Program.[150] Exposures to 0, 50, or 100 ppm EO for 6 hr/day, 5 days/week began in August 1981. No report has yet been issued.[97]

VIII. EVALUATION AND CONCLUSIONS

A. Local Sarcomas and Gastric Lesions

By virtue of its chemical reactivity, EO is a highly irritant material capable of causing injury to skin, mucous membranes and other tissues of the body.[151] It comes as no surprise, therefore, to find that EO is sarcomagenic at the site of repeated subcutaneous injections in mice,[152] and that repeated gavage with EO in oil elicits reactive changes and tumors in the forestomach of rats, and to a lesser extent in the glandular stomach. The part played by repeated tissue injury at these subcutaneous or intragastric sites is uncertain; systematic studies of the pathogenesis of the lesions have not been carried out. However, the background information set out below should provide some perspective in this situation.

B. Tumors Associated With EO Irritancy

Highly irritant materials like EO induce a local inflammatory response in the tissues with which they come in contact. If the response is permitted to regress before some crucial point in time, by removing the stimulus, progression to neoplasia does not take place. On the other hand, continuation of repeated application of the irritant material tends to prevent the healing process, so that progression to tumor formation is often the long-term outcome.[152-157] If histopathology is not performed early in the study, the inflammatory response may go undetected, and epithelial hyperplasia, cellular dysplasia or neoplasia will be seen in the later stages of the process. The point should be emphasized that the phenomenon is caused by high concentrations of the chemical irritant at the site of application. A potent chemical carcinogen, which is highly soluble, and readily absorbed and distributed, would be expected to induce tumors at sites distant from the local site of application. Secondary tumors or metastases to remote sites may be present. These are more dependent on the nature of the primary tumor than upon the agent that induced it. Therefore, in terms of human hazard assessment, the absence of metastases at remote sites is of less importance than the absence of primary neoplasms at remote sites.

C. Tumors of the Forestomach

Dunkelberg[144] demonstrated the induction of tumors of the forestomach in rats treated twice weekly with EO in oil. The tumors spread through the gastrointestinal system and appeared late in the course of the study. Tumors did not arise in organs or tissues remote from the site of application. If EO were a potent carcinogen specifically for the gastric mucosa, then one would expect that the gastric reaction product of EO and HCl, 2-chloroethanol, would also be carcinogenic. Dunkelberg[159] tested 2-chloroethanol using a protocol like the one used for evaluating EO and the results were negative in spite of the use of very large doses. It is reasonable to conclude that the irritant properties of EO are necessary to induce hyperplasia leading to tumors of the forestomach upon repeated application.

Forestomach tumors in rats spread by their very nature. Stomach tumors in man are quite prone to metastasize. Therefore, the presence of metastases of this type of tumor in experimental animals is more an indication of the metastatic potential of the tumor itself than an indication of the carcinogenic potential of EO. The significance of the observation that there were no primary neoplasms induced at remote sites by intragastric administration of EO derives from the facts that large doses were given repeatedly, EO is readily absorbed and distributed, and large tumor yields at other sites would have been expected if EO were indeed a potent carcinogen. Since no remote tumors were observed, the action of EO on the forestomach may well be due to its irritant local effects in high concentrations.

A recent review of the method employed by Dunkelberg[144] has recommended caution in the use or interpretation of gavage studies using an oil solvent or suspending agent. The report states:

> "Over recent decades, an increasing body of scientific information has accumulated to suggest that the practice of administering test substances in edible oils to laboratory animals may introduce confounding factors which could affect the accuracy, the relevance and interpretation of the experiment insofar as considerations of human health are concerned." Ad Hoc Working Group on Oil/Gavage in Toxicology.[160]

Mechanistic factors considered by the Ad Hoc Working Group which might apply to Dunkelbergs' study include: the effect of oil composition (including toxic oxidation products of polyunsaturated fatty acids) on membrane composition and response to chemical and endocrine stimuli, lipid peroxidation and its relationship to tumor promotion or cocarcinogenesis, and modification of the immune response. Interpretation of studies like that by Dunkelberg[144] is complicated by the uncertainty of these factors, but most of all by the local impact of massive and repeated doses of a highly reactive chemical agent dissolved in oil.

There is an analogy here with the case of butylated hydroxyanisole (BHA), which also elicits tumors of the forestomach in the rat.[149,149a] Nera et al.[149b] have demonstrated that BHA induces an early (9 days) irritant and proliferative response in the epithelium of the forestomach. Similar studies have not been done with EO.

D. Subcutaneous Sarcomas

A potent chemical carcinogen, when given subcutaneously, should induce tumors at the site early in the study, should do so at low doses, should not induce an inflammatory response, and may cause tumors at other sites. Dunkelberg[143] gave weekly injections of EO in mice and produced sarcomas at the site of injection, no tumors at remote sites, used very large doses, and found the tumors to be late-occurring. It is impossible to establish whether acute and/or chronic inflammation was involved, since there were no interim kills to study early stages in the pathogenesis of the disease process. The significance of the sarcomas in the subcutis is questionable for human hazard assessment.[157a]

A plan has been proposed for determining the relative carcinogenic potency of a chemical based on the induction of subcutaneous sarcomas.[152] Application of the scheme to the data by Dunkelberg,[143] suggests that EO is a substance "acting only indirectly" (e.g., a modulator of the spontaneous incidence) or at most, a "weak carcinogen". This same reference discusses a number of chemicals which have been evaluated in chronic subcutaneous studies, including iron dextran. This material is listed in the National Toxicology Program Second Annual Report on Carcinogens[161] as a potent inducer of local sarcomas in experimental animals, and yet has been widely used in human and veterinary medicine for 30 years without a single authenticated instance of associated subcutaneous sarcoma.[162] In the case of iron dextran, the route of administration is the same in man as in the experimental animals. In the Dunkelberg study, EO was given by an inappropriate route of exposure. Experience with a variety of materials, evaluated for their carcinogenic potential using this technique, illustrates the caution which should be exercised in estimating the carcinogenic potency of any chemical by repeated subcutaneous injection.[152-157a]

Tomatis[163] reported only infrequent false positive and negative results (14.3% of 102 chemicals) when the subcutaneous route was used as a predictor of carcinogenic potential demonstrated by other routes for the 222 chemicals evaluated by the International Agency for Research on Cancer. This tabulation was in part inspired by a letter to the Editor, *Journal of the National Cancer Institute* by A. F. Pelfrene.[164] Pelfrene criticized a subcutaneous injection study published in that journal and provided arguments for abandoning this approach to carcinogenicity assessment, citing a host of individual scientific authorities and research organizations. These arguments were rebutted by the study authors and later by Tomatis.[165] The original criticisms by Pelfrene and his rebuttal of Tomatis' response emphasize the widely held position that many noncarcinogenic materials result in formation of injection site sarcomas upon repeated subcutaneous injection. Results from this screening technique are not, therefore, amenable to human carcinogenic hazard assessment.

IX. HYPERPLASIA ASSOCIATED WITH EO IRRITANCY OR POSSIBLE INFECTION

A. Nasal Metaplasia

Squamous metaplasia of the nasal respiratory epithelium may be caused by repeated exposure to high concentrations of an irritant, and it is reversible up to some critical point in time. When the irritant is removed before that time, the change reverses within a few months. Therefore, nasal metaplasia itself does not necessarily commit an experimental animal to development of nasal malignancy. In addition, where severe rhinitis of infectious or allergic origin is present, squamous metaplasia is a concomitant pathological change, both in man and animals.

Nasal hyperplasia and metaplasia were reported in the NIOSH chronic rat inhalation study on EO. These changes may have been due in part to the irritant effects of EO. Such changes were not seen in the BRRC study, and the only difference in exposure was that the NIOSH rats were exposed for 1 hr per day longer. A possible additional explanation is found in the pathology report from the NIOSH study:[122a]

> "The rats exposed to 50 ppm and 100 ppm of Ethylene Oxide had a higher incidence of inflammatory lesions of the lungs, nasal cavities, trachea, and internal ear. The lungs of these rats had an increased incidence of bronchiectasis and bronchial epithelial hyperplasia, and a greater severity of inflammatory changes. The inflammatory lesions of the lungs consisted of various degrees of multifocal suppurative bronchitis and bronchopneumonia, abscess formation, hemorrhage, edema, macrophage accumulations, and an increased number of peribronchial and perivascular mononuclear cells. The internal ear of many of these rats had a unilateral or bilateral suppurative otitis media. The changes in the nasal cavity consisted of varying degrees of suppurative rhinitis, hyperplasia of the respiratory epithelium, and multifocal areas of squamous metaplasia of respiratory epithelium. All of these lesions were similar to the severe manifestations seen in the chronic respiratory diseases complex in rodents."

This suggests that the *Mycoplasma pulmonis* infection may have been largely responsible for the fulminating infection and accompanying squamous metaplasia in the nasal cavity and may have compromised the immune status of the animals.

Clarification is needed of the interpretation of the study done on hydrochloric acid (HCl) at New York University, which is often cited as evidence that squamous metaplasia induced in rats by inhalation of formaldehyde is not associated with its irritant effects.[158] HCl at 10 ppm is not irritating to the nasal mucosa of the rat, as indicated by the absence of histological changes in the nasal epithelium. The RD_{50} for HCl is 309 ppm, as opposed to 3.13 ppm for formaldehyde.[166] High levels (e.g., at least 30 or 100 ppm) of inhaled HCl would be required to achieve an irritant effect. Therefore, HCl at 10 ppm does not cause squamous metaplasia and does not cause nasal tumor formation while formaldehyde at 14.3 ppm does both.

B. Adrenal Cortical Hyperplasia

Proliferation of cells in the adrenal cortex was reported in the NIOSH study. The likely cause of this observation was suggested by the above quotation (describing the extreme state of infection and consequent disease present in all rats) which preceded the following statement:

> "The rats exposed to 50 ppm and 100 ppm of Ethylene
> Oxide had a high incidence of proliferative and degenera-
> tive lesions of the adrenal cortex. These changes were of
> slightly greater incidence and severity in the rats exposed
> to 100 ppm of Ethylene Oxide. The changes consisted of
> vacuolation and hyperplasia or hypertrophy of the cells of
> the zona fascicularis. In some of these rats the areas of
> hyperplasia and hypertrophy formed distinct nodules
> which compressed the adjacent cortical tissue. These le-
> sions are called cortical nodular hyperplasia in the Histo-
> pathology Incidence Tables."

It is clear that the likely cause of the adrenal cortical hyperplasia was infection with
Mycoplasma.

A high incidence of endocrine disturbances associated with the development of endocrine tumors is characteristic of long-term rodent studies in which animals are fed
ad libitum.[167] This age- and diet-dependent stress may have contributed to the adrenal
cortical hyperplasia in the NIOSH rats. In the BRRC rats, which did not display histologic lesions associated with the chronic respiratory disease complex of rodents, adrenal changes were limited to fatty infiltration of the cortical region.

X. INHALATION STUDIES

For purposes of evaluation, a detailed analysis of various facets of the BRRC and
NIOSH studies is presented in Appendix B. In addition, Dr. R. L. Sielken has provided
a study of the implications of the time to response information (see Appendix C).

EO has been shown to increase the incidence of tumors, or decrease the latency
period for tumor formation, in at least one chronic animal study carried out by inhalation, the relevant route of exposure. The tumors which have been observed have all
been late-occurring neoplasms which occur spontaneously in the Fischer 344 rat. No
unique tumors were produced by EO exposure, suggesting the possibility that EO is
active through a mechanism which does not involve initiation but rather promotion or
another form of modulation of the spontaneous tendency to tumor development.

The results which are best able to support a quantitative assessment of human cancer
risk are those of the BRRC study, with a preference expressed for use of the incidence
data on peritoneal mesotheliomas. The leukemia data, while capable of supporting a
quantitative risk assessment, offer a lower level of confidence in the results because of
the higher spontaneous incidence of MCL than peritoneal mesothelioma in Fischer 344
rats,[168] the possible specificity of this disease for this strain of rat, and the late-occurring nature of the disease.

The findings on pituitary adenomas provide striking supportive evidence for a role
of EO as a modulator of spontaneously occurring neoplasia.   Several features of the
gliomas induced in the BRRC and NIOSH studies are important with respect to their
impact on human hazard assessment, to be discussed in Chapter 11. First, the gliomas
are less common in F344 rats than other tumors whose incidence was increased by EO
exposure. Second, the primary brain neoplasms may have more direct relevance to
human brain neoplasms than rat MCL has for the vast majority of human leukemias.
Finally, the appearance of gliomas very late in the lifespan of the exposed rats indicates
that EO does not affect the latency of gliomas, just as it does not decrease the latency
of other spontaneous rat tumors.

The term "glioma" encompasses a number of pathological types of primary brain
tumor. Whether they can all be lumped together for purposes of hazard assessment is

questionable. The Report of the Ad Hoc Panel[148] provides guidelines for combining such data; while it permits all gliomas (i.e., oligodendrogliomas and astrocytomas) to be combined, it does not allow gliomas to be combined with granular cell neoplasms, nor with nerve cell neoplasms. This rule has often not been observed in the course of human hazard assessment (for instance see Table 4, Chapter 11).

Chapter 9

INDUSTRIAL HYGIENE ASPECTS

I. INTRODUCTION

Industrial hygiene airborne guidelines and standards have been continuously developed to control worker exposure. In 1951 the Manufacturing Chemists Association recommended a maximum workplace EO exposure limit of 100 ppm over 8 hr.[170] The Factory Mutual Company Engineering Division recommended (date unknown) the following EO exposure guidelines as "probably safe".[171]

Daily exposures up to 7 hr duration:	50 ppm
Occasional (2 weeks — repeated exposure up to 4 hr/day):	100 ppm
Single exposures of several hours' duration (up to 7, no more than 1/week):	150 ppm
Repeated exposures of up to 1 hr/day:	150 ppm
Single (no more than 1/week) exposure of up to 1 hr duration:	500 ppm

Inhalation toxicity studies by Hollingsworth et al.[63] and Jacobson et al.[62] prompted Hine and Rowe in 1973 to recommend permissible exposure limits of 100 ppm for 4-hr daily exposures over a 2-week period and 50 ppm for 7-hr daily exposures.[76]

In 1971 the Occupational Safety and Health Administration (OSHA) adopted as a federal regulation the 1968 American Conference of Governmental Industrial Hygienists (ACGIH) Threshold Limit Value (TLV) of 50 ppm EO as an 8-hr (TWA.) OSHA lowered the workplace standard to 1 ppm TWA, in a final rule published in June 1984 (see Preface).

II. PERSONAL MONITORING

The measurement of EO vapor in workplace air can be achieved by a variety of methods. Odor is not a good indicator of EO at low concentrations because the odor threshold in humans is approximately 700 ppm. Olfactory fatigue has been reported at high concentrations.[55]

Accurate measurement of EO vapor requires the use of instrumental detection techniques. Personal or area samples may be gathered over a period of time with a collection device for subsequent laboratory analysis. Sampling yields an integrated concentration over the sampled period. Alternatively, area measurements may be taken in the workplace using direct reading instruments.

Early EO air sampling methods relied upon fixed or semiportable sampling pumps and equipment. Only area or "breathing zone" samples were gathered with reagent filled impingers, silica gel trap tubes, or evacuated flasks. The analytical methods for these samples employed wet chemical techniques which exhibited wide variability and difficulty with accurate detection below 10 ppm. Accurate measurements down to 5 ppm were reported using impinger collection and hydration to ethylene glycol, oxidation to formaldehyde, and colorimetric determination following reaction with sodium chromotrope.[172]

With the advent of portable battery-powered air sampling pumps in the 1960s, personal air sampling for EO became possible. Sampling and analytical methodologies for EO were developed and improved, permitting lower detectable limits to be achieved. The air sampling methods currently in common use are charcoal tubes and passive dosimeters. Other methods, infrequently employed, include sampling bags and impingers.

The charcoal tube sampling technique is used most widely for personal EO monitoring. Several different charcoal tube sampling methods are available. They differ primarily in the amount and type of charcoal used in the collection tubes.[173] The most widely used charcoal tube technique is the Qazi-Ketcham method which permits a lower detectable limit of 0.15 ppm EO in a 10 *l* air sample volume.[174] A recent refined version of the charcoal tube technique has shown a lower detectable limit of 0.03 ppm EO, and has been validated thoroughly in the range of 0.3 to 20 ppm.[175]

A variation on the charcoal tube technique has been developed by OSHA. A reliable lower detection limit of 0.052 ppm is claimed for this method which uses a standard 150 mg sKc charcoal tube to collect a 1 *l* air sample in 20 min.[176] In contrast to data collected by OSHA, general industrial experience and tests conducted by industrial hygiene laboratories indicate that the standard sKc charcoal tube has poor EO collection and storage capability. Field validation data for the OSHA method have not been reported to date.

In general, the advantages of charcoal tube sampling methods are that small, portable sample collection tubes are used, good sensitivity can be achieved, and interfering airborne substances are usually not a problem.[173] The disadvantages are as follows: sample loss from charcoal "channeling" or overloading can occur from poor technique, results are suspect in areas of high temperature or humidity, short sampling periods ($<$ 15 min) are not possible, and sample storage and shipment must be carried out under cold conditions.

Impinger methods for personal EO sampling are available but are not widely used. One of the impinger methods has been validated in the 1 to 8 ppm range with 94.2% EO recovery.[177] The advantages of the impinger method are that collection efficiency is not subject to heat and humidity, a wide range of flow rates and sample volumes can be used, and samples can be stored and shipped at room temperature. Disadvantages arise from the fact that the impinger is cumbersome for personal sampling, handling of the liquid may result in sample loss, and the impinger must be weighed before and after sampling to correct for evaporative water loss.[177]

Monitoring with gas sampling bags consitutes a third personal monitoring method. The lower limit of detection, accuracy, and precision of analysis are dependent upon the type of gas chromatograph detector and the technique used. The advantages of bag sampling are that it can be used for long- or short-term sampling, no sample loss is observed at 50 ppm for a 5-day storage period using Tedlar bags (polyvinyl fluoride), and the bags are reusable. On the other hand, the bags are bulky to use, transport, and store, and they may be punctured by sharp objects.[173]

There are two types of passive dosimeters for personal EO monitoring currently undergoing development and validation, including field validation and testing. Because of their sheer simplicity (no pumps, hoses, calibration, etc.), these devices are likely to revolutionize EO sampling when fully validated under field conditions.

In sum, the general advantages of personal monitoring are portability; sensitivity; specificity for EO; allows calculation of a time weighted average exposure; and measures actual personal exposure. Disadvantages of the method are laboratory analysis necessary; time delay for results; and results do not identify peak emission sources or events related to exposure.

III. BIOLOGICAL MONITORING

Biological monitoring to measure the effects of EO exposure is not in widespread use. While certain indexes exist, such as the production of SCEs or alkylation of hemoglobin, none of these are specific for EO. Although cytogenetic endpoints may bear a relationship to the extent of exposure to EO, conclusions cannot be drawn as to the biological significance of these observations.

IV. AREA SAMPLING

Area sampling of EO may be accomplished by a variety of direct reading instruments. These devices are able to make quantitative determinations of vapor concentrations in the immediate vicinity of the instruments' detector probe. As such, they are suitable for area measurements of ambient air concentrations. They are not suited for full-shift personal exposure monitoring applications.[178]

Direct reading instruments for EO may be portable or fixed in nature. In the former instance they are useful for leak detection and to pinpoint short-term levels or peak emission points. In the latter case they may be useful continuous monitors, often having sensor points at several locations within the workplace. Alarms may be fitted to most monitors and set to activate at a predetermined level for emergency evacuation purposes or other reasons.

A wide variability exists between the capabilities of available direct reading instruments with respect to specificity, detection limits, and response time when used for EO. Of these factors, specificity is the most significant in determining a particular instrument's utility for EO measurement.[173] False readings will result if the nonspecific instrument measures an interfering compound and this is mistaken for EO. Choice of instrumentation is highly dependent upon the potential interferences present and the lower detection limit required. Lower detection limits are in the range of 0.1 to 1.0 ppm. These devices are not generally suitable for measurement of personal time-weighted average exposures.

V. EO PRODUCTION AND USE

The major part of the EO produced in the U.S. is used in closed-system chemical manufacturing processes, a practice dictated by the flammability and reactivity hazards associated with the chemical. This method facilitates a close degree of control over routine worker exposures within established permissible limits in the producer and converter industries.

In 1976 about 0.02% of the U.S. EO production (or 500,000 kg) was used in health care facilities for sterilization purposes.[76] Under these conditions of use, the control of worker exposures is somewhat more difficult owing to the greater potential for physical contact of workers with EO gas from sterilizer operations and freshly sterilized materials.

VI. PRODUCER/ETHOXYLATOR INDUSTRIES

Engineering changes in the facilities of producer/converters of EO have been evolutionary in nature and often related to productivity improvements and concern about explosibility and flammability. These changes, however, have resulted in the reduction of potential employee exposures.

In general, facility modifications have been oriented to:

1. The collection of plant streams for environmental, product recovery and/or energy conservation purposes.
2. Increasing degrees of automation which have reduced the number of manual operations and which have allowed staffing with fewer operating personnel.
3. Economies of scale, which have resulted in fewer facilities and thus reduced the number of employees potentially exposed.

A. Control Measures

Specific engineering efforts in production/conversion units which have resulted in decreasing potential exposures include the installation of double mechanical pump seals, double seated valves, check valves, plugs on vents and drains, flanged leak detectors, returns for process vents to combustion or recovery systems, remote isolation of EO pumps, and exhaust hoods at sampling points.

These and other similar changes have resulted in employee exposures which are generally well below 5 ppm in the production/conversion units. Somewhat higher potential exposures exist for maintenance and repair of process equipment and during quality control sampling.

Engineering changes have also been implemented in overland distribution facilities to reduce potential employee exposure. Such changes include the installation of sample hoods, routing sample hood drains back into the process, improvements in filter change techniques, special thin valves for railcar host disconnections and hot water and nitrogen purging of lines prior to breaking. Potential employee exposures in distribution operations are much more difficult to control than in production/conversion units because of the need to continually break lines and connections for railcars and tank trucks. Because of this, respiratory protection is generally used for critical distribution jobs.

VII. HEALTH CARE INDUSTRY

Significant advances have been made in the health care industry to reduce employee exposure to EO below the existing 50 ppm TWA OSHA standard. One of the more challenging areas is the evaluation and control of EO levels emitted from freshly sterilized materials, for in most cases there is a need for further handling of these materials during subsequent manufacture and distribution. The potential for chronic low-level exposure to the workforce necessitates this evaluation.

One published report on the use of small sterilizers states

> "the material removed from the sterilizer at the end of the cycle contains EO dissolved in certain materials, adsorbed onto others, contained within air spaces of packaging, etc. This material releases EO to the surrounding atmosphere, and can act as an extended source of EO in the workplace. Without control, chronic exposures of between 5 and 40 ppm can occur, depending upon the type and amount of sterilized material, and the size of the specific workplace where it is stored and handled."[179]

The amount of EO and rate of release from sterilized goods is extremely variable and dependent upon their composition. Some items (e.g., metal surgical instruments) will degas instantaneously. Other items, such as surgical gloves, gauzes, and corrugate, will release varying amounts of EO over long time periods. Consequently, large multiple pallet sterilizers containing a large volume of certain materials may give rise to even higher exposure levels.

A. Control Measures

Use of ventilated areas for off gassing of freshly sterilized materials can effectively reduce these exposures. Small sterilizers may require only a cabinet, whereas large multiple pallet operations require a special room. Keeping the area under negative pressure, ventilation to outside the workplace (no recirculation) and control of employee access using proper safeguards are all important. For the large sterilizer degas rooms, massive ventilation is required to help reduce employee exposures during transfer of goods or removal of bio-indicators (or other quality control samples). The latter operation poses a special problem because the employees may receive short-term high exposures when opening containers to remove quality control samples. Control measures, in addition to dilution ventilation, must be instituted to protect these employees.

Another point in health care operations where EO emissions are known to occur is when sterilizers are opened after purge cycles have been completed. Exhaust ventilation is often the control method of choice. Local exhaust ventilation is generally preferred over dilution ventilation because it provides a more positive degree of control and requires smaller air volumes. Some sterilizers can be equipped with back-draft valves as a form of local exhaust ventilation.

In principle, use of properly designed respirators affords complete protection against ambient EO concentrations. For purposes of risk assessment, consideration must be given to actual exposures, i.e., are they the same as ambient concentrations if respiratory protection is provided?

Standard cartridge-type respirators offer insignificant protection against EO. Airline or self-contained breathing apparatus do provide respiratory protection but the equipment tends to restrict the mobility of the worker. Less expensive, practical respirators are being developed which include canister units with passive end-of-service-life indicators (ESLI). In a July 1984 Federal Register Notice, NIOSH announced that it would accept applications for approval of respirators with active or passive ESLI. Criteria were set forth for NIOSH approval.[179a]

Administrative scheduling of work location/duration to limit EO exposure can be effective in certain situations, providing that a proper, continuous management system is used to control personnel movements. Implementation of such a program requires that the time relationships of exposure levels be well known in areas of potential exposure. An effective air monitoring program is important. In order to exercise effective control over workforce exposures, the employer must understand and act upon the relationship between measured levels, time spent in areas, and full-shift exposure.

Chapter 10

EPIDEMIOLOGIC STUDIES

I. INTRODUCTION

Concern has been generated by recent reports suggesting that EO may be a human carcinogen and teratogen. This concern is largely based on data derived from generally well-controlled animal experiments, along with a limited number of reports and studies of occupational exposure. Although the purpose of this Chapter is the consideration of evidence derived from human observations, it is important to be aware of the rather extensive bibliography reporting the mutagenic and clastogenic properties of this substance, as reviewed in Chapter 7. Evidence of carcinogenicity is available from a 2-year chronic inhalation study which linked EO to the induction of peritoneal mesotheliomas in male rats and mononuclear cell leukemia in female rats[27] (see Chapter 8). These observations were confirmed in an independent NIOSH study on rats. Both studies show associations with CNS tumors.[27a,b] Evidence that EO is teratogenic in laboratory animals is virtually nonexistent. Some effects have been reported at the highest exposure levels, but these are generally thought to be related to maternal toxicity.

This chapter is intended to provide an overview of available information concerning chronic health effects in human populations exposed to EO. As mentioned in the Introduction and discussed in Chapter 9, human exposures to EO occur predominantly in two settings: the chemical production process, characterized by relatively low average exposures and infrequent accidental exposures; and the sterilization process, characterized by relatively higher average exposures and sporadic peak exposures (anticipated at regular intervals in the process). The reports to be discussed here come from each of these settings. A critical analysis of each report is presented in Appendix A.

The problem of concomitant exposures is endemic to most epidemiologic studies. Nonetheless, in the case of EO, some mention of this issue is in order. Kaye[180] has stressed that, in Sweden, exposure results from the use for sterilization of 50% EO in methyl formate, a mixture whose toxicity to, and reactivity with, human cells cannot be equaled to that of EO alone. In one of the Swedish studies[74] exposure to a variety of chemicals, in addition to EO, was involved.

The purpose of this Chapter is not to attempt an integration of observations between animals and man but rather to summarize the general information and insight provided by the occupational studies and reports that have been published to date.

II. CARCINOGENICITY

Direct evidence does not exist that will permit a conclusion that EO is a *known* human carcinogen. Suggestions of human carcinogenicity are available from four reports (Table 1); one[75] from the U.S., two from Sweden[74,181] and one from Germany.[182] One of the Högstedt et al. reports[181] was not a study because no real attempt was made to identify a cohort nor to trace their health experience over time. Rather, this is a case report of three employees who developed blood dyscrasias following exposure at a sterilization plant: one case of chronic myeloid leukemia, another of acute myelogenous leukemia and a third of macroglobulinemia, which is a rare hemopoietic malignancy, not generally classified as a "leukemia". The authors estimated that only 230 persons had ever worked in exposed areas, and from these only 0.2 cases of leukemia would have been expected. However, as with all reported clusters, these numbers do

Table 1
STUDIES OF CANCER INCIDENCE AND
MORTALITY ASSOCIATED WITH EO
EXPOSURE

Cohort size (exposed)	Setting	Exposure		Ref.
		Average	Peak	
175	Production	< 20	700 +	181
767	Production	< 10	700 +	75
(230)	Sterilization	—	—	74
602	Production	< 10	≤ 2,000	182

not lend themselves to rigorous statistical analysis and are of little use in evaluating the specific risks attributable to EO. In addition, the sterilization gas used was a 50-50 mixture of EO and methyl formate, so that workers were exposed to this toxic material as well.

The other three studies were conducted in factories that *produced* EO. The Högstedt et al. study[74] examined mortality experience of 89 workers exposed directly in the production area, 86 men exposed intermittently during maintenance activities, and 66 men who had never worked in the exposure area. All workers had survived a minimum 10-year latent period following first exposure at the plant. Average exposures at this plant were estimated to have ranged between 5 to 20 ppm, with occasional peaks at or above the odor threshold of 700 ppm. The Morgan et al. study[75] examined the mortality experience of 767 workers potentially exposed during the production of EO in a Texaco plant where time weighted average exposures were estimated to be less than 10 ppm. Although not stated, it would seem reasonable to assume that occasional "high" peaks also occurred at this plant. The Thiess et al. study[182] examined the mortality experience of 602 employees exposed to alkylene oxides and their derivatives between 1928 and 1980 at eight German production plants. The average operation period was 14 years and over half survived a minimum 10-year observation period. Exposures (based on post 1968 data) were generally less than 10 ppm with excursions to 2000 ppm.

These studies generated conflicting results (Table 2). The Hogstedt et al. study[74] found a significant excess of deaths attributable to "all malignant neoplasms", "cancer of the stomach", "leukemia", and "all circulatory diseases" among the directly exposed workers. There were 9 cancer deaths in the direct group vs. 3.4 expected, 3 stomach cancers vs. 0.40 expected, and 2 leukemias vs. 0.14 expected. Unfortunately, the small cohort size (89 persons and 1324 person-years in the direct groups) and the small number of deaths which occurred limit the interpretation of these findings. No significant excess mortality was found for the intermittently exposed group. Of the 3 cancer deaths observed in the intermittent group, there was 1 stomach cancer vs. 0.4 expected and 1 leukemia death vs. 0.1 expected.

The second cohort study, reported by Morgan et al.,[75] was based on a larger number of exposed workers (767 persons and 13,969 person-years) all of whom had a minimum of 5 years' exposure. This study yielded negative results. Compared with 15 expected, 11 cancer deaths occurred in this cohort. No cases of leukemia or stomach cancer were found. The authors did not state the number of workers surviving a minimum 10 year latency period, which would have enabled better comparison of results with the Högstedt study.

The authors of the third study[182] concluded that further evaluation was needed, but interpreted the results from their study as indicating no increase from cancer or other causes of death in this cohort of workers exposed to alkylene oxides. In the total co-

Table 2
CANCER INCIDENCE AND MORTALITY STUDIES:
RESULTS

	Observed	Expected	SMR 95% confidence limits	Ref.
Group 1: Direct Exposure (89)				
Cancer Mortality	9	3.4	1.2 —5.03	74
Leukemia	2	0.14	1.6 —51.6	
Stomach	3	0.40	1.51—21.91	
Cancer Incidence	11	5.9	0.93— 3.34	
Group 2: Intermittent Exposure (86)				
Cancer Mortality	3	3.4	0.18— 2.58	
Stomach	1	0.4	0.03—13.91	
Leukemia	1	0.1	0.13—55.64	
Group 3: Non-exposed (66)				
Cancer Mortality	1	2.0	0.01— 2.78	
Stomach	0	—		
Leukemia	0	—		
Potential Exposure (767)				
Cancer Mortality	11	15.24	0.36— 129	75
Leukemia	0	0.70	0— 5.24	
Stomach	0	—		
Brain	2	0.70	0.35—10.31	
Total Cohort (602)				
Cancer Mortality	14	16.65	0.46— 1.41	182
Leukemia	1	—		
Stomach	4	2.67	0.41— 3.84	
Brain	1	—		
10-Year Latency (351)				
Cancer Mortality	11	12.57	0.44— 1.57	
Leukemia	1	0.145—0.148	0.17—38.02	
Stomach	2	1.76—1.85	0.13— 4.00	
Brain	1	0.066—0.071	0.37—81.31	

hort, there were 14 observed (O) vs. 16.6 expected (E) cancer deaths. For the subcohort with the minimum observation period of 10 years, the numbers were 11 (O) and 12.6 (E). SMRs for leukemia (0=1), stomach cancer (0=4), and brain cancer (0=1) were greater than 1.0 but not statistically significant, nor was there evidence that they were occupationally related based on length of employment and latency analyses.

Because of the probable similarity of production plant exposures, it is not likely that the differences in results of these three cohort studies are attributable to differences in EO exposure levels. Except for the possibility that average exposures were slightly lower (less than 10 ppm) in the Morgan[75] and Thiess[182] cohorts, typical exposure patterns could be expected to be similar. However, exposures to other chemicals, including at least one suspect carcinogen [bis (2-chloroethyl) ether] were more prevalent in the

Table 3
REPORTED LEUKOCYTE FINDINGS (1000/mm³)[a]

	Exposed (N)[c]	Control (N)	Ref.
Total leukocytes	9.12 (40)	7.55 (41)	77, 183
Lymphocytes[b]			
Exposure Status			
Permanently exposed	2.19 (17)		
Previously exposed	2.04 (32)	2.02 (54)	
Intermittently exposed	2.13 (52)		

[a]   Other hematological changes are discussed in Chapter 8.
[b]   Analysis restricted to "healthy" workers.
[c]   Number of persons.

Högstedt cohort since EO was produced by the chlorohydrin process. This could conceivably have caused some of the observed difference in cancer incidence reported.

Although another report by Joyner[77] is frequently cited as a negative cancer study, the design of that investigation did not include any follow-up of exposed workers and therefore no evidence of carcinogenicity or other chronic health effects could be expected.

No other prospective studies of EO workers have been reported. Lacking any strong evidence of a dose response relationship in the study by Högstedt et al.,[74] nor any confirmation of their findings in other exposed populations, an assertion of the existence of a direct cause and effect relationship cannot be supported at this time.

III. DISCUSSION

The suggestion of possible chronic health effects associated with EO that arises from the reports of Högstedt et al.[74,181] should be balanced against the weaknesses and uncertainties of the data. Overall, the conclusion is reached that both the quantity and quality of the available epidemiology studies are inadequate to permit any sound decisions to be made regarding quantitative health risks for exposed workers. A case can be made, however, for the contention that the epidemiological evidence does provide an upper limit for the likely risks that may have existed in the past.

IV. OTHER HEALTH EFFECTS

A. Lymphocytosis

The suggestion has been made that acute high level exposures to EO may stimulate total leukocyte or lymphocyte production (Table 3). This finding is based on reports by Ehrenberg and Hällström[183] and by Joyner.[77] Although in both studies an elevated white cell or lymphocyte count was documented for exposed EO workers, it should be noted that these counts were within the normal range. Therefore this finding is of unknown significance and no evidence of any associated ill health has been reported. In addition, there is some indication in the Ehrenberg and Hällström study that this elevation, if related to EO, was transient. A reduction in lymphocyte counts in 13 exposed workers was found after exposure levels were reduced through improved ventilation of factory air.

Further information is forthcoming from two more recent studies of employees potentially exposed to EO in the course of manufacturing the gas, and/or using EO in the production of other chemical agents.[75a,b] Both studies involved potential exposure levels generally below 10 ppm (8-hr TWA), often below 1 ppm, and even below the detection limit of the personal air samplers (0.05 ppm) during table EO manufacturing operations. Neither study revealed a significant difference from the corresponding unexposed control group with respect to any hematological indices, nor any departure from the normal limits of control populations. Van Sittert et al.[75a] did find a positive correlation between duration of employment in EO manufacturing and the percentage of neutrophils in a differential leukocyte count; there was a negative correlation with the percentage of lymphocytes in the blood of potentially exposed workers. The percentage and absolute numbers of B and T lymphocytes in whole blood were not different in plant workers and controls but tended to be somewhat higher in smokers than nonsmokers. The only unexpected finding by Currier et al.[75b] was an increased prevalence of proteinuria in their EO group.

B. Reproductive Effects

The possibility of decreased sperm counts associated with EO exposure has been addressed in one report by Abrahams.[184] The author concluded that no valid interpretation of the results could be made because of the insufficient number of sperm samples obtained.

In a nationwide study of 80 hospitals throughout Finland, Hemminki et al.[86] found an approximate twofold increase in the incidence of spontaneous abortions among workers exposed to EO. An age-adjusted spontaneous abortion rate of 12.7% was reported among 146 exposed pregnancies compared with a rate of only 7.7% among 1004 unexposed pregnancies. However, problems with the study design and statistical analysis, which have been identified and discussed by Austin[276] and Gordon and Meinhardt,[277] together with the absence of any other studies on this subject, limit the conclusions that can be drawn from this study. Studies of spontaneous abortion are particularly difficult to conduct due to the fact that many early abortions are unreported and difficult to diagnose accurately. Because this study did not restrict its focus to hospitalized abortions, it relied heavily on the individual's ability to recall, diagnose, and accurately record their reproductive history on self-administered questionnaires. Because those responsible for selecting study subjects, as well as the subjects themselves, were aware of the study hypothesis regarding EO, there is some possibility that bias could have been introduced. In addition, it is possible that the authors did not completely control for the more accurate recall of spontaneous abortions associated with the most recent pregnancies, which tended to be the exposed ones. Another major problem in analyzing data on "pregnancies" is that the requirement of "independent events" is violated when more than one pregnancy per woman is analyzed, as was true in this study. Although the analysis of confirmed hospitalized cases tended to support an increased rate for EO exposed pregnancies, it was based on too few pregnancies, i.e. 31, to permit meaningful conclusions to be drawn.

A more detailed analysis of the report by Hemminki et al.[86] is provided in Appendix A.

C. Cardiovascular Diseases

Evidence of increased mortality from cardiovascular disease was reported by Högstedt et al.;[74] however, Thiess et al.[182] found evidence of significantly decreased cardiovascular mortality. These significant, but conflicting findings, suggest that the results were artifactual and unrelated to EO exposure.

V. CYTOGENETIC STUDIES

A. Chromosomal Aberrations

A relatively larger body of information is available on the frequency of chromo-

Table 4
RESULTS OF CYTOGENETIC STUDIES IN WORKERS EXPOSED TO EO

Subgroup	Chromosomal aberrations per 100 Cells			
	Exposed (N)		Control (N)	Ref.
Maintenance workers	17.5 (7)		4.3 (10)	183
Sterilizer operators	4.5—8.8 (75)		3.6 (41)	184
Long-term exposure (> 20 year)	3.5 (11)			187
Short-term exposure (< 20 year)	2.33 (6)		1.0—1.4 (25)	
Long-term acute exposures	2.23 (21)			
Short-term acute exposures	1.60 (5)			
Packers[a]	6.0 (12)		5.3 (11)	116
Sterilizers[b]	6.8 (5)			
High	1.5 (2)[c]	1.1 (24)[d]	0.6 (11)	186
Moderate	0.7 (4)[c]	0.9 (18)[d]	0.5 (19)	
Low	— (8)[c]	0.4 (5)[d]	0.7 (21)	

[a] 0.5 to 1.0 ppm, 8 hr/day, 5 days/week for 0.8 to 8 years.
[b] 5 to 10 ppm for 1 hr/day.
[c] High potential exposure employees.
[d] Low potential exposure employees.

Table 5
EFFECT OF EO EXPOSURE ON INCIDENCE OF SCEs

Nature of exposure	SCEs (mean no. per cell)			
	Exposed (N)		Control (N)	Ref.
Chronic (symptomatic)	9.7 (4)			73
Chronic (asymptomatic)	8.6 (8)		6.4 (8)	
Incidental	9.4 (12)			
	8.0—9.2 (75)		5.4 (41)	184
	(H)[a]	(L)[b]		
High	32.3 (2)	14.7 (24)	11.4 (22)	186
Moderate	14.8 (4)	10.2 (18)	10.8 (19)	
Low	10.0 (8)	10.5 (5)	8.8 (12)	
High	10.7 (9)		7.6 (13)	192
				259
Low	7.8 (5)			
Total	8.8 (14)			
Median	8.2 (14)		7.5 (13)	

[a] High potential exposure employees.
[b] Low potential exposure employees.

somal aberrations in cultured lymphocytes of EO exposed workers. These studies include both numerical and structural aberrations, as well as SCEs (Tables 4 and 5). The results of these studies are difficult to interpret because of the unknown connection between these changes and any specific type of illness. Moreover, the background rate of gross numerical and structural chromosomal aberrations is approximately 2/100 cells and other agents (therapeutic X-rays, smoking, caffeine, viruses, and environmental pollutants) are known also to increase aberration frequencies.[185] Good studies, therefore, are particularly difficult and expensive to conduct. Special care must be

Table 6
SUMMARY OF CYTOGENETIC STUDIES IN WORKERS EXPOSED TO EO

Occupational setting	No. of exposed groups	EO exposure (ppm)		Ref.
		Ave. or range	Peak	
EO production	1	"High"	(Accidental)	183
Sterilization	1	50	75	
EO production	4	5.0	1900	182
Sterilization	2	0.5—10.0		116
	3	5—200		186
	2	36	1500	73
	2	High or low[a]		192
Sterilization and packing	2	<1.0[b] <0.5[b,c]	5.0	121a
Sterilization	2	10.7±4.9[b,d] 0.35±0.12[b,d]		121b
EO production	1	< 0.05	8.0	75a,n
Sterilization	6	L:0.5[b] M:5—10[b] H:50—200[b,e] 5—20[b,e]		75e
Sterlization and related tasks	1	14—113		75f,g

Note: L,M,H = low, moderate, and high exposures.

[a] Exposures expressed as "estimated cumulative dose of EO" during 6 months preceding cytogenetic studies: high exposure, > 100 mg EO, group mean 501 mg; low exposure, < 100 mg EO, group mean 13 mg.
[b] TWA(8 hr).
[c] Two of the 10 subjects were exposed to 1 to 2 ppm (TWA$_8$).
[d] Mean ± SD.
[e] Corrective action reduced the TWA$_8$ from 50 to 200 down to 5 to 20.

exercised over every detail of technical performance, including the concurrent analysis of lymphocytes from control subjects.[185a]

As Table 6 indicates, at least 12 cytogenetic studies have been reported recently, 3 from production and 9 from sterilization facilities. Of these studies, eight included two or more exposure groups, but only the Johnson and Johnson[186] study was able to look at dose-response relationships to some extent. Information on peak exposures in the study populations was not available in all cases. For instance, the Ehrenberg and Hällström[183] study was limited to seven workers accidentally exposed to extremely high levels.

When aberration frequencies (excluding SCEs) between exposed and control workers are compared within five of the studies which reported these data, one sees a consistent gradient of difference, with higher rates found in exposed groups. With the exception of one study,[183] these differences are not large. However, one also sees a wide range of reported values for "control", presumably nonexposed, employees. This wide range of variability is partially attributable to the fact that the Johnson and Johnson study[186] reported only "complex aberrations" (exchanges). However, without the Johnson and Johnson data, values range from 1.0 to 5.3, overlapping with the values reported in two[184,187] of the remaining four studies. This could reflect variation in procedures used at different laboratories, real differences among control groups, or nonsimultaneous testing.[185a,188]

Little evidence of a dose-response relationship can be found in these data. This fact is due, in part, to the manner in which exposure groups were formed. In the Thiess et al.[187] study, for example, differences relate to "length of exposure" and whether or not an accidental exposure occurred — not to some continuum of dose. In the Pero et al.[116] study, packers were exposed "continuously" to low levels whereas sterilizers were "intermittently" exposed to slightly higher levels. In the Johnson and Johnson[186] study, workers in three exposed groups were studied in three plants where exposure took place, but no dose-response relationship was seen due to small numbers of observations and small differences in exposure rates.

In summary, there is reason to question the interpretation of an EO "effect" in these studies because of overlap in the aberration values reported for control and exposed workers, lack of a dose-response relationship, the small number of workers studied and the small differences reported. However, when all these studies are taken together, there is some suggestion that, at *high-levels* (50 ppm or more), EO is capable of increasing chromosomal aberrations. At low-level exposures (10 ppm or less), no effect on chromosomal aberrations is manifest.

B. SCEs

SCEs have been examined in studies of exposed workers (Table 5). In addition, Hansen et al.,[189] in a study of 14 hospital sterilizer workers exposed to < 5 ppm TWA$_s$ EO (< 50 ppm 15 min peak), found no significant difference in mean SCE levels, as compared with matched controls. Because of the much higher natural background frequency of these SCE changes (6 to 8 per cell), fewer cells per worker are required to detect significant increases. The results of these studies indicate a tendency of highly exposed EO workers to have higher SCE rates than controls — in spite of the small number of workers examined, the relatively small changes observed, and the slight overlap of reported values for exposed and control series.

C. Reversibility of Cytogenetic Changes

Of these studies four attempted to decide whether elevated aberration frequencies in exposed individuals returned to normal over time (Table 7). Despite the possibility that the altered frequencies obtained at retest were due to sampling variation, differences in laboratory techniques or natural background variation, the persistence of elevated SCE frequencies in instances of high exposure to EO is a striking and intriguing feature of the more recent data.

In summary, available evidence suggests that exposure to high levels of EO can increase the frequency of chromosomal aberrations and SCEs; however, there is little certainty regarding correlation of these changes with significant health end-points (see below).

D. SCEs: a Health Effect of EO?

Exposure to EO is unusual in that it takes two more or less distinct forms: the fairly steady level characteristic of production or ethoxylation, and described by a TWA; and the recurrent exposure interspersed by peaks and excursions as exemplified by those that may occur during sterilizing operations using EO. The ability of EO to reach and react with nucleophilic sites in DNA is likely to be very different under these two circumstances. Reaction with competing nucleophiles en route to nuclear DNA, as well as the impact of competing cellular detoxication mechanisms, will probably combine to minimize the amount of DNA alkylation that occurs during fairly constant, low-level exposure. DNA repair is likely to be rapid, so that — especially after cessation of exposure — the integrity of the DNA will be restored.[121c] The apparently tight coupling of SCE formation to DNA synthesis means that if the lesions that give rise to

Table 7

FOLLOW-UP MEASUREMENTS OF SCEs AND CHROMOSOMAL
ABERRATIONS

Nature of exposure		SCEs/Cell (mean)		
		Initial	Retest	Ref.
Chronic (symptomatic)	(N = 4)	9.79	10.3	73
Control	(N = 8)	6.37	—	75e,186

	Worksite exposure[a]	Follow-up period (months)			
		Initial (N)	6 (N)	12 (N)	24 (N)
High potential exposure	High	32.3 (2)	35.2 (2)	21.4 (2)	21.1 (2)
	Moderate	14.8 (4)	10.3 (4)	14.9 (4)	12.5 (4)
	Low	10.0 (8)	—	9.5 (8)	10.5 (8)
Low potential exposure	High	14.7 (24)	15.1 (24)	13.5 (23)	12.5 (22)
	Moderate	10.2 (18)	9.1 (16)	9.9 (18)	10.3 (15)
	Low	10.5 (5)	—	9.3 (5)	10.8 (5)
Unexposed	High	11.4 (22)	12.2 (23)[a]	11.8 (20)	11.8 (20)[b]
	Moderate	10.8 (19)	8.7 (19)	10.1 (21)	10.2 (18)
	Low	8.8 (12)	—	9.6 (13)	10.0 (11)

Chromosomal Aberrations per 100 Cells (mean)

		Initial	Retest	184
Exposed	(N = 75)	1.14—1.83	0.80—3.00	
Long-term	(N = 11)	3.5	2.7	187

[a] High and low potential exposure employees compared in three exposure level settings (plants); "high" plant 5 to 200 ppm, "moderate" plant 5 to 10 ppm, and "low" plant 0.5 ppm, TWA_8.

[b] Community controls: 8.7 (29), 9.8 (28).

SCEs are repaired before DNA synthesis occurs, SCEs will not be formed during the subsequent S phase.[190] One may speculate that the occurrence of peaks of EO exposure, even though they may be brief, is more likely to introduce a surge of EO that will effectively alkylate DNA, to a degree that is not readily reversible before the next cell division "locks in" the abnormality as a mutation.[191]

These two different modes of exposure represent extremes that are not, perhaps, a complete reflection of real-life situations. Yet the distinction is important in the context of available evidence of genotoxicity of EO, and particularly the fact that exposure above certain levels elicits the formation of SCEs in peripheral lymphocytes of man and animals.[121,192] The dose-response relationship of SCE formation, and their persistence after exposure has ceased, has been interpreted as evidence of a cumulative action of EO, expressed as an integrated dose rate over the period of exposure. A need exists for more information to relate both steady states and peak effects of EO exposure in terms of the corresponding SCE levels in lymphocytes. Preston[193] has considered the validity of applying data such as those of Yager et al.[192] and Yager and Benz[121] for calculating TWA_8 with respect to human exposure. He points out that the first need is for "a clear-cut multipoint dose-response curve". The data of Yager et al.[192] are totally unsuitable for this purpose. Whether in men or rabbits, the conditions involved brief, intermittent, acute exposures to EO; they cannot be translated to chronic 8 hr/day, 5 days/week exposures in any appropriate manner.

There is mounting evidence that the level of SCEs in human peripheral lymphocytes is a useful indicator of exposure to EO at levels equal to or exceeding 10 ppm and that measurements of SCEs have proved of considerable value for purposes of biological monitoring of exposed individuals.[185,194-196] However, SCEs can be induced by a variety of chemicals, both in vitro and in vivo[118,197] and the incidence can be influenced by a number of factors such as concentration of BUdR,[198-200] growth temperature,[201,202] and culture medium composition.[203-205] Variations in interpersonal response in different reference populations and the DNA repair capacity of the cell limit the usefulness of the assay for monitoring populations exposed to chronic low levels of EO.[206]

From the standpoint of hazard assessment, consideration should be given to the question whether a rise in SCE level is indicative of a health effect.[207] In order to answer this question an understanding is required of the mechanism(s) by which SCEs arise. A strong correlation is known to exist between alkylation at the O^6 position of guanine and the induction of forward mutations in mammalian cells.[208,209] Alkylation at the O^6 position of guanine was suggested as a major factor in SCE induction[210-213] when it became clear that the extent of overall DNA binding by an alkylating agent was not strictly related to the induction of SCEs.[214] However, this view is becoming increasingly untenable. Painter[215] suggested a replication model of SCE formation, mediated by the enzymatic machinery at the replicating fork.[216] Carrano and Thompson[217] demonstrated that the efficiency of SCE formation relative to the induction of mutations differed with each agent tested.

It has become apparent that there is no clear correlation between SCE formation and unrepaired O^6-alkylguanine adducts.[214] Indeed Morris et al.[218] have gone so far as to say that SCEs may not be related to mutations at all, but to reduction in cell survival. In CHO cells in culture, where mutations are induced at the HGPRT locus, the alkylation-induced damage that leads to SCE formation is related to the damage causing cell death to a greater degree than to mutations. These same cells have provided an illustration of the distinction between extensive chromosomal aberrations produced by methylene chloride (with or without metabolic activation), simultaneously with an unchanged level of SCEs.[219]

The question of the relationship of chromosomal aberrations to SCEs is of considerable interest. In an analysis of the published findings for 215 compounds, Gebhart[75f] found a lack of correspondence between the results of the two cytogenetic tests in 30% of the tested materials. Comparison of the SCE data with those for point mutations yielded similar results. Stetka et al.[75c] tested the consistency of the relationship between induced SCEs and induced mutations, using CHO cells exposed to ethylnitrosourea (ENU) and then held at confluency for various times; periodic assays for SCEs and for mutations were carried out. The frequencies of induced SCEs underwent an exponential decline with holding time, whereas the mutation frequencies remained constant. The conclusion drawn was that, for ENU under the conditions of the experiment, SCEs and mutations may represent manifestations of different forms of DNA damage. Of those forms of DNA damage induced by ENU, at least some that elicit SCEs are repaired, and therefore differ strikingly from the forms of DNA damage that lead to mutations and are not repaired so readily. A considerable number of publications have now appeared, dealing with the distinction between induction of SCEs and of mutations, together with the possible basis for these different manifestations of genotoxicity in terms of different types of primary or secondary DNA damage.[75i-l] The dependence on class of genotoxic agent used is illustrated by the work of Pal[75m] on SCEs formed in CHO cells. He found that a close parallel existed in the course of DNA repair, between reduction in frequency of SCEs and decrease in the level of deoxyribonucleoside-diol-epoxide adducts produced by *anti*-benzo(a)pyrene-7,8-diol 9,10-oxide. The striking capacity to repair SCE-associated DNA damage caused by EO was demonstrated in human peripheral blood lymphocytes by Hedner et al.,[121c] but no concurrent measurements of changes in DNA adducts have yet been carried out.

Thus, while a definitive mechanism of SCE induction is not yet at hand, the current trend may be interpreted to indicate that SCE induction and mutagenesis are separate and perhaps independent events. An increase in SCE levels may be viewed as an indicator of exposure to various physical or chemical agents, for example, exposure to peroxidizing unsaturated fatty acids,[260a] ultrasound,[261,262] or in the course of smoking tobacco products,[220] but should not be considered in hazard assessment of EO, save at exceptionally high levels of exposure.

Chapter 11

HAZARD ASSESSMENT

I. INTRODUCTION

Scientific risk assessment is a critical component of the process in which regulatory agencies reach social judgment decisions regarding exposure to chemicals identified as potential health hazards. The function of scientific risk assessment is to provide the regulator with a foundation upon which to make management decisions on public health issues. Decisions of this nature require a balancing of the risk associated with use of the product, placement of the defined risk within the context of other risks acceptable to society, and consideration of costs associated with limiting exposure.

The component parts of a scientific risk assessment are as follows.

1. Hazard identification, i.e., qualitative identification of a substance posing a potential risk.
2. Hazard assessment, i.e., evaluation of all valid biologic, toxicologic and epidemiologic data relevant to an assessment of human risk.
3. Exposure assessment — quantification of exposure and characterization of the exposed population.
4. Characterization of the risk to the exposed population as quantitatively as may be feasible, based on consideration of 1 to 3.

The foregoing sections of this document have provided a comprehensive and critical analysis of the hazards associated with exposure to EO. Quantification of exposure and characterization of the exposed population are not included — hence the characterization of the document as a "hazard assessment". The purpose of this chapter is to draw conclusions concerning the most important biological effects of EO, based as far as possible on their dose-response relationships, and to apply this information for purposes of hazard assessment for human exposure to EO. There will be three approaches considered: first, the mathematical extrapolation that has now become traditional; second, the concept of "radequivalence"; and third, an assessment based on a biological perspective of the totality of the available evidence on EO.

II. USE OF MATHEMATICAL MODELS FOR QUANTITATIVE RISK ASSESSMENT

There has been considerable discussion in the literature on the use of mathematical dose response models to predict potential risk at exposure levels below the experimental base. The discussion has centered largely on the use and selection of models to extrapolate from high-dose experimental data to the low doses to which humans are generally exposed, and on the criteria or formulas for cross species extrapolation to human risk from animal data.

In the case of EO, both the BRRC study[27,27a,b] and the NIOSH investigation[122,122a] were designed to evaluate the tumorigenic potential of EO in rats exposed to well-defined chronic exposure levels. These investigations produced dose-response data which can be used for quantitative risk assessment; however, there are facets of these studies which raise questions concerning their use as predictors of cancer risk in human populations through the application of mathematical models.

A detailed discussion of issues relating to the BRRC and NIOSH studies may be found in Appendix B. The major considerations bearing on use of these studies for risk assessment purposes are as follows.

First, the extent to which epizootic infections (SDA virus in the BRRC rats and *Mycoplasma* infections in the NIOSH study) influenced the final outcome of the study merits careful consideration. Interaction, if any, between the infectious agent and the test material would necessarily complicate interpretation of the ensuing experimental results.

Second, the relevance to man of the tumorigenic effects observed in F344 rats is uncertain. The best-defined tumorigenic effect in both the BRRC and NIOSH Fischer 344 rat studies is enhancement of the incidence of mononuclear cell leukemia (MCL), a type of leukemia with a high spontaneous incidence in the Fischer 344 rat but not, for example, in the Sprague-Dawley rat or other mammals. The relationship, if any, between MCL of F344 rats and human forms of leukemia is discussed in Appendix B. The kinds of leukemias that were purported to be associated with exposure to EO in humans[181] differed among the three reported cases, and were quite distinct from mononuclear cell leukemia, both in the expression of the neoplasia and its course. Another carcinogenic endpoint observed in the animal studies — peritoneal mesothelioma — while clinically a form of cancer that is found in man, has not been associated with exposure to EO in epidemiology studies. The appearance of primary brain neoplasms in both the BRRC and NIOSH investigations is of interest — at least at first glance. While the relevance of these tumors for man may be greater than in the case of MCL or PM, the data satisfy only one (the dose-response relationship) of the six criteria for neurocarcinogenic action (see Appendix B). Hence it seems inappropriate to use these results in quantitative risk assessment.

Third, conclusions of the BRRC study, and presumably the NIOSH study as well, are complicated by the absence of a unique early-occurring tumor. Differences in the incidence of spontaneous tumors in senile rats do not carry the same weight in assessment of carcinogenicity as unique tumors occurring earlier in life. In addition, the wide variation in the background incidence of spontaneous tumors in the F344 rat, particularly MCL, makes interpretation difficult. It is interesting to note that after up to 26 months of exposure at the highest dose level (100 ppm), the leukemia incidence in male rats exposed in the Bushy Run study was lower than the reported spontaneous incidence in control rats in the NIOSH study.

Nevertheless, these chronic inhalation studies are the only investigations employing a route of administration relevant to human exposure in which a dose-response was observed for carcinogenic end-points. Thus, the experimental results can be applied to mathematical models and the results of that treatment of the data are set out below.

The assessment will be based on the BRRC data plus the male mononuclear cell leukemia data from the NIOSH study.[122,122a] Tabulated data relevant to the quantitative risk assessment are presented in Tables 1 to 21.

A. Animal Exposure Regimens and Worker Experience

Before the data can be applied to a model, differences between the exposure regimens in the animal studies and the human worker experience must be accommodated. (There is also a slight difference between the exposures in the two laboratory tests — 6 hr/day at BRRC and 7 hr/day at NIOSH, for 25 and 24 months, respectively.) The exposures of human interest are for 8 hr/day for a fraction of a lifetime. In order to reconcile these differences, a "normalizing" procedure is introduced whereby each exposure (dose) is converted to its "continuous lifetime equivalent", as if every breath contained this level for an entire lifetime. For example, animals exposed to 1 ppm, 6 hr/day 5 day/week have a continuous lifetime equivalent of 0.179 ppm, where:

Table 1
MCL IN FEMALE RATS: BRRC STUDY[260]

Period (months)	Experimental dose (0 ppm)			
	Died on test		Killed	
	With response	Without response	With response	Without response
0—3	0	0		
3—6	0	0	0	20
6—9	0	0		
9—12	0	0	0	20
12—15	0	8		
15—18	1	3	0	40
18—21	4	7		
21—25	6	11	11	104

$$0.179 = 1 \times \frac{6}{24} \times \frac{5}{7}$$

If animals were exposed for two thirds of their lifetime to 1 ppm for 7 hr/day, 5day/ week, they would have a continuous lifetime equivalent of

$$0.134 = 1 \times 2/3 \times 7/24 \times 5/7$$

By this same normalizing procedure, a worker with a lifetime expectancy of 73 years, employed from, for example, age 20 through age 60 to a TWA_8 of 1 ppm of EO, 5 day/week, 48 week/year, would have a continuous lifetime equivalent of

$$0.120 = 1 \times 40/73 \times 8/24 \times 5/7 \times 48/52$$

Such a procedure is strictly a mathematical convenience to permit intercomparison of data, e.g., the BRRC Study (where daily exposure was for 6 hr), the NIOSH study (where exposure time per day was 7 hr) and a possible workplace situation in which a person may be potentially exposed to EO for 15, 30, or 45 years.

Use of the continuous lifetime equivalent in the present treatment, while convenient for normalizing different test protocols and equilibrating widely variable human exposure conditions, has definite limitations and drawbacks. It is probably useful, and fairly accurate, to employ the concept for some degree of chronic exposure to fairly low levels of EO. For example, the concept may adapt well to a person exposed daily for 15 years at concentrations of the order of magnitude of 25 ppm, or for 40 years (possibly with interruptions) at 5 ppm, and the like. But the concept surely cannot be used for sporadic, or episodic, exposures of short duration at relatively high levels such as those encountered in some sterilizing operations.

Although supportive data are lacking, consideration of the reactivity of EO, its solubility, demonstrated ability to be converted, metabolized, and excreted from biological systems, leads to the conclusion that it is scientifically impossible to equate quantitative cumulative exposures, from episodic short-term bursts, with those exposures which might result from continuous low-level concentrations. Thus, it is extremely doubtful that any conclusions, which may be drawn from quantitative risk assessment based on chronic studies and employing the "normalization" procedure described above, are applicable to short-term exposure situations.

Table 2
MCL IN FEMALE RATS: BRRC STUDY[260]

Period (months)	Died on test		Killed	
	With response	Without response	With response	Without response
0—3	0	0		
3—6	0	0	0	10*
6—9	0	0		
9—12	0	0	0	10*
12—15	0	2		
15—18	0	4	0	20*
18—21	1	4		
21—25	2	9	11	43

* Indicates an outcome from an early sacrifice where the microscopic examination was made only if the gross examination showed a lesion. The assumption that the absence of a gross lesion implies the absence of MCL has no effect on the estimated proportion of responders corrected for early deaths (0.2438) or on the SE (0.0568).

It is also important to note that in establishing continuous lifetime equivalents, absorption/dissemination/elimination parameters between species (including man) are not taken into account, and it is assumed that, in any one species, the amount of EO absorbed or translocated to a potential target site varies directly with the number of months per years exposed. Data to develop an evaluation of such factors are lacking.

B. Choice of a Model(s)

The basic procedure for selection of a dose-response model or models requires statistical examination of how well the model fits the experimental data. Where more than one model fits the data, the basic criteria for selection and weighting results are how good the fit is and the reasonableness of the assumptions underlying the models in relation to the known data.

Choosing a mathematical dose-response function (model) is an important step in a risk assessment. Numerous models have been proposed; all leading contenders are mathematical generalizations of the one-hit model. These models may give very different risk assessments when the data set to which they are applied is dissimilar from a one-hit model situation. However, the one-hit model provides an excellent fit to the five data sets for EO (Table 14). Therefore, only the one-hit model will be used here initially.

$$P = 1 - e^{-(\alpha + \beta d)}$$

where: d = the dose (continuous lifetime equivalent); P = the proportion of responding animals; and α, β = parameters to be estimated from the data. After estimating α and β with a data set, one can estimate the excess risk ΔP as*

* ΔP is obtained as follows. Take P in the equation $P = 1 - e^{-(\alpha + \beta d)}$. If $d = 0$ (no dose, therefore background), then $P_0 = 1 - e^{-\alpha}$; ΔP, then, is obtained from Abbots formula, in which

$$\Delta P = \frac{P - P_0}{1 - P_0}, \text{ or } \Delta P = \frac{1 - e^{-(\alpha + \beta d)} - 1 + e^{-\alpha}}{1 - 1 + e^{-\alpha}},$$

which reduces to $\Delta P = 1 - e^{-\beta d}$.

Table 3
MCL IN FEMALE RATS: BRRC STUDY[260]

| | Experimental dose (33 ppm) | | | |
| | Died on test | | Killed | |
Period (months)	With response	Without response	With response	Without response
0—3	0	0		
3—6	0	1	0	10*
6—9	0	0		
9—12	0	0	0	10*
12—15	0	5		
15—18	1	3	0	20*
18—21	4	5		
21—25	5	7	14	34

* Numbers are "early deaths" (0.4053) or on the SE (0.0644).

$$\Delta P = 1 - e^{-\beta d}$$

The statistical procedure of weighted least squares will be used here to fit this model to a data set. The equation may be expressed in linear form as

$$Y = -\ln(1 - P) = \alpha + \beta d$$

The weight w for each datum is

$$w = \frac{(1 - \text{obs. } P)^2}{(\text{SE of obs. } P)^2}$$

Table 14 gives the results of fitting the one-hit model to each of the five data sets for EO. Notice that the one-hit model fits well in every case as evidenced by the large *P*-values. The first data set (female rats with MCL) is the worst case, because the estimated value of β is largest.

If one considers an industrial worker exposed to EO by inhalation, e.g., to an exposure of 10 ppm, TWA$_8$ over a 30-year working experience (assuming 48 working weeks per year, 5 day/week), this person's continuous lifetime equivalent dose is 0.92 ppm, where:

$$0.92 = 10 \times \frac{8}{24} \times \frac{5}{7} \times \frac{48}{52} \times \frac{30}{72}$$

In Table 15 this (second row, second column) and other lifetime equivalent doses for possible working experiences are presented.

The next step is to estimate the excess cancer risk (ΔP) for a worker exposed at a continuous lifetime dose (d). As mentioned above, the formula

$$\Delta P = 1 - e^{-\beta d}$$

in Table 16 gives calculated results (all five data sets) for various assumed lifetime continuous doses. Thus, the worker in the previous paragraph had a continuous lifetime dose of about 1.0 ppm (0.92). Using this approach and this model, the data would

Table 4

MCL IN FEMALE RATS: BRRC STUDY[260]

| | Experimental dose (100 ppm) | | | |
| | Died on test | | Killed | |
Period (months)	With response	Without response	With response	Without response
0—3	0	0		
3—6	0	0	0	10
6—9	0	0		
9—12	0	2	0	10
12—15	0	13		
15—18	0	6	1	19
18—21	2	5		
21—25	10	9	15	11

predict an excess cancer risk (ΔP), the magnitude of which is dependent on the particular endpoint in the BRRC or NIOSH animals studies used as the basis for the assessment. Note that four endpoints predict an excess risk of 10^{-2} while the predicted risk from the PBN data is approximately one order of magnitude less.

As a second example, the case of a worker exposed to EO by inhalation over 15, 30, or 45 years of a working lifetime (based a lifetime expectancy of 72 years) to 1.0 ppm TWA$_8$, 5 day/week, and 48 week/year would, by a similar treatment, lead to the predicted excess risk shown in Table 17.

One can compare the excess predicted risk from the above treatment with that derived by OSHA, as published in the Notice of Proposed Rulemaking.[2] In the OSHA treatment, data from the BRRC study were used for female rats (MCL) and male rats (PM). The OSHA data have been "translated" from the listing as excess lifetime risk per 10,000 workers, and the data given are for the "Maximum Likelihood of Excess Risk" (MLE) data only, for both the multistage and one-hit models (Table 18). For purposes of comparison, the parameters of exposure conditions for humans are essentially equivalent.

It is apparent in these comparisons that the estimates of excess risk are essentially of the same magnitude or differ by up to circa three-fold. The OSHA method utilizes a "scaling factor" for transferring the treatment dose to humans on a milligram per kilogram basis assuming 100% absorption rate for both animals and man. The EOIC approach utilizes the "continuous lifetime equivalent" outlined above. Further, the numbers used as "responders" taken as a proportion of those "at risk" differ in the two treatments, especially for the female mononuclear cell leukemia data. In view of the uncertainties outlined elsewhere in this chapter regarding translation of observed animal effects to calculation of risks to man, in addition to the lack of information regarding comparative pharmacokinetics, it is surprising that the results of these different approaches are as similar as they are.

Other assumptions which make these approaches essentially "worst-case" treatments are

1. Man's exposure for less than a lifetime, postmaturity, is carcinogenically equivalent to the rat's lifetime exposure.
2. There is no universal threshold on the dose response curve below which the excess risk from exposure to EO is zero.
3. The excess risk at low dose levels is linear with respect to exposure concentration.

Table 5
ESTIMATED PROPORTION OF RESPONDERS IN
TABLES 1 TO 4 (CORRECTED FOR EARLY
DEATHS) AND SE OF THE ESTIMATE[260]

Level	Proportion responders	SE
0 ppm actual in test (0 ppm equivalent lifetime dose)	0.1704	0.0332
10 ppm actual in test (1.79 equivalent lifetime dose)	0.2438	0.0568
33 ppm actual in test (5.89 equivalent lifetime dose)	0.4053	0.0644
100 ppm actual in test (17.86 equivalent lifetime dose)	0.7074	0.0741

The statistical results of a further exploration of the latter two assumptions are presented in Table 21. First, the possibility of a threshold may be considered. The historical approach is to test the response of each treated group to the response from the appropriate control group. From the data in set 1, the four doses (0, 1.79, etc.), the four proportions of responding animals (0.1704, 0.2468, etc.), and their four standard errors (0.0332, 0.0568, etc.) are reproduced (Table 21). Finally, the one-sided p-value for comparing each treated group to its control is given (0.13 for 1.79 vs. 0, etc.).

The one-sided p-values appear to reach significance (< 0.05) at continuous lifetime doses between 1.79 and 5.89 for the BRRC data sets (1 to 4) and at about 10.42 for the NIOSH study (data set 5). These continuous lifetime doses correspond to administered dose levels of 10 and 33 ppm, and 100 ppm, in these studies, respectively. The statistics indicate the possibility of a threshold on the dose-response curve, above a continuous lifetime dose of 1.79 ppm, below which the excess risk from exposure is zero.

Second, the linearity of excess risk at low dose levels is considered. One alternative to the one-hit model with its low-dose linearity is the Weibull model with a shape parameter m as an exponent for the dose. The risk at a low dose (d) is proportional to the dose raised to this power (d^m). Thus, if m = 1, the Weibull model reduces to the one-hit model. If m is greater than 1, then the Weibull model gives smaller risks than a method based on assumed low-dose linearity. If m is less than 1, the Weibull model gives larger risks than a method based on assumed low-dose linearity. The best estimate of m is given for each of the five data sets in Table 21. In no case is the best estimate significantly different from m = 1, or even close to being significantly different. It appears, therefore, that the Weibull model reduces to the one-hit model and linearity at low dose levels is predicted.

In addition to estimating the lifetime risk of a tumor, it is also possible to estimate the time at which the tumor develops. The serial kills from the BRRC study provide the additional information needed to make this estimate. The usual mathematical model for this is

$$P = 1 - e^{-(\alpha + \beta d)t^k}$$

where: t = the time to a tumor; and k = the shape parameter for the time. Other symbols are as previously defined. The tumors associated with EO exposure appear very late in life, and only the 18-month results are of any value in estimating the value of the shape parameter k. Table 19 shows the estimated value of k for animals which developed MCL in the BRRC study. The risk at 18 months was calculated according to the life-table method. In four of the eight groups, the estimated risk at 18 months is zero, and the estimated value of k is infinite. The other four estimated values of k are

Table 6
MCL IN MALE RATS: BRRC STUDY[260]

	Experimental dose (0 ppm)			
	Died on test		Killed	
Period (months)	With response	Without response	With response	Without response
0—3	0	0		
3—6	0	1	0	20
6—9	0	0		
9—12	0	0	0	20
12—15	0	3		
15—18	2	4	0	40
18—21	5	4		
21—25	18	19	13	84

near ten, which is a very large number for this shape parameter. This means that these tumors appear *very* late in life (a k of about 6.5 is more typical for carcinogens.)[221] Among the animals which *do* develop an MCL from exposure to EO, the average (median) animal develops MCL at 93.3% of its normal lifetime — or at 23.3 months of life (= 0.933 × 25).

The time-to-tumor aspect of the BRRC study has also been dealt with at length by Sielken[222] (see Appendix C). He has stressed the lateness of the neoplastic response, in both male and female rats, with respect to both MCL and PM, particularly at the lower exposure levels. Using the following criteria, the 10 ppm results were virtually identical with those of the unexposed controls: duration of exposure required to elicit MCL, or PM, in 5% of the males or females; mean time-to-tumor, or average duration of freedom from MCL, or PM, in males or females; percentage of 25-month survival, combining all risks, in either males or females. There was a very slight difference in these parameters when rats exposed to 33 ppm were compared with control groups. Thus, only the 100 ppm animals showed an effect.

Finally, one can consider the question of relating the risk assessments based on the animal results to available human epidemiologic data. Högstedt and his colleagues[74] have followed a cohort of 87 male Swedish workers exposed to EO.* Each worker had at least 1 year of exposure and at least 10 years of induction-latency. There were 1324 person/years observed for an average of 15.2 years/person (= 1324/87). As of 1977, significant findings in this cohort were 2 deaths from leukemia vs. 0.14 expected and 3 deaths from stomach cancer vs. 0.40 expected.

It is necessary to make some specific numerical assumptions in order to complete the calculations. According to Högstedt et al.,[74] the expected total number of deaths for the cohort is 13.5, or 16% (= 13.5/87). Using this figure and the death rates for Swedish males, one estimates that the average age of the cohort in 1977 was about 63 years. Högstedt refers to exposure as early as 1941; arbitrarily, we can assume 25 years of exposure as an average for the cohort. Högstedt mentions 25 mg/m³ (14 ppm) as the typical exposure level for this average worker. Thus, the average worker had, in 1977, a continuous lifetime equivalent exposure of 1.22 ppm, where

$$1.22 = 14 \times \frac{8}{24} \times \frac{5}{7} \times \frac{48}{52} \times \frac{25}{63}$$

* There were 89 persons included in the exposed group; however, Högstedt dropped two from the analysis because they had less than 1 year of exposure.

Table 7
MCL IN MALE RATS: BRRC STUDY[260]

Period (months)	Died on test — With response	Died on test — Without response	Killed — With response	Killed — Without response
0—3	0	0		
3—6	0	0	0	10*
6—9	0	1		
9—12	0	0	0	10*
12—15	0	0		
15—18	0	2	0	20*
18—21	3	3		
21—25	9	9	9	42

* Indicates an outcome from an early sacrifice where the microscopic examination was made only if the gross examination showed a lesion. The assumption that the absence of a gross lesion implies the absence of MCL has no effect on the estimated proportion of responders corrected for early deaths (0.2438) or on the SE (0.0590).

With this continuous lifetime exposure, the implied excess risk according to the results from the animal experiments is

$$\Delta P = 1 - e^{-\beta \times 1.22 \times \left(\frac{63}{73}\right)^k}$$

where: β = best estimates (Table 14); $k = 10$ (Table 19) for the animals in the Bushy Run study; and $k = 6.5$ is a typical value for other human cancers. Table 20 gives the outcome of this calculation for the five animal data sets and for both values of k.

Finally, the observed excess risk from the Högstedt study is calculated. For leukemia as an end-point, it is

$$\Delta P = \frac{2.0 - 0.14}{87 - 0.14} = .021$$

where 2.0 is the observed number of deaths and 0.14 is the expected number. This observed excess risk should be compared to the calculated excess risks in Table 20.

Ultimately, this cohort of 87 male Swedish workers will live out its expected lifetime. Because the excess leukemias in rats were significant late in life, one can predict that there should be some more leukemia deaths among these workers, again based on the assumption that the rat is a predictor for man. In 1987, the average member of the cohort will reach his expected lifespan of 73 years. If he has zero exposure after 1977, then his lifetime equivalent exposure in 1987 will be 1.05 ppm, where

$$1.05 = 14 \times \frac{8}{24} \times \frac{5}{7} \times \frac{48}{52} \times \frac{25}{73}$$

The excess risk (ΔP) will then be

$$\Delta P = 1 - e^{-1.05\beta}$$

Table 8

MCL IN MALE RATS: BRRC STUDY[260]

	Experimental dose (33 ppm)			
	Died on test		Killed	
Period (months)	With response	Without response	With response	Without response
0—3	0	0		
3—6	0	0	0	10[a]
6—9	0	1		
9—12	0	2	0	10[a]
12—15	2	2		
15—18	1	2	0	20[a]
18—21	3	5		
21—25	7	14	12	27

[a]   Numbers are "early deaths (0.4592) or on the SE (0.0694)".

where β comes from the animal study. As an illustration, consider β = .0576 (data set 1). Then, the excess risk is ΔP = 0.059 (vs. 0.016 for k = 10 or 0.027 for k = 6.5 in 1977).

Admitting to all the uncertainties attendant upon attempts to translate the observed animal health effects to possible risk in humans, two other observations concerning the outcome of the mathematical treatment of the data are in order.

First, the high tumor incidence in humans exposed to EO, as predicted by application of models to the animal data, has not been verified in epidemiological studies, i.e., a strikingly increased incidence of one or more types of cancer in workers has not been observed. It may be argued that some epidemiology studies have not been pursued for a sufficient length of time, or the cohorts sufficiently large, to permit observation of the predicted outcome, especially in light of the long latency periods associated with the neoplasms that might result. If such is the case, the obviously correct course of action is to continue to observe human populations where there has been significant exposure, and to quantify this exposure, even if only retrospectively.

Second, for a potential tumorigenic agent, such as EO, where the expression of its effects experimentally is that of a low-level enhancement of spontaneous tumor types in the particular species of animal under investigation, the experimental lifetime exposure of animals is not adequate for prediction of risk to humans. Indeed, this is borne out by the results of the statistical approach used in the "time-to-tumor" calculation, in which a modest effect is seen that translates, even superficially, to only a brief life-shortening effect. Unfortunately, the validity of this conclusion suffers from the fact that the only "time-to-tumor" data are derived from the BRRC study, the protocol for which was not designed to provide adequate numbers of animals at the interim levels for valid statistical time-to-tumor evaluation.

III. RADEQUIVALENCE: AN APPROACH TO RISK ASSESSMENT

Another approach to quantitative risk assessment is based on the concept of "radequivalence". A risk assessment relating the degree of hemoglobin alkylation in EO-exposed workers and that caused by units (rads) of γ-radiation has been proposed by Ehrenberg and others (see below). The idea that EO exercised a radiomimetic effect on individuals exposed to its vapor was put forward by Ehrenberg and Hällström,[183] who compared chromosomal aberrations induced by EO to the action of total body γ-irra-

Table 9
MCL IN MALE RATS: BRRC STUDY[260]

Period (months)	Experimental dose (100 ppm)			
	Died on test		Killed	
	With response	Without response	With response	Without response
0—3	0	0		
3—6	0	0	0	10
6—9	0	0		
9—12	0	1	0	10
12—15	0	6		
15—18	0	3	0	20
18—21	4	7		
21—25	13	15	9	21

diation. Bridges[223,224] suggested the expression of probability of genetic damage caused by a chemical mutagen (under conditions of environmental exposure) in terms of "radequivalent," i.e., the dose of x- or γ-radiation which would produce the same effect. The concept of "radequivalence" has been used repeatedly for purposes of risk assessment by Ehrenberg and his colleagues[3,6,21,22,24,95,225,226] in spite of serious reservations and debate regarding the validity of the approach.

Committee 17 of the Environmental Mutagen Society[227] proposed that, in order to estimate the mutagenic risk to man from exposure to chemical agents, based upon data from a variety of assay systems, it was clearly advantageous to employ a basic unit for chemical effects, whereby use could be made of the large amount of data already available for radiation. They suggested the use of remequivalent-chemical (REC). However, the applicability of the REC was discussed to some extent in the report of Committee 17, and was also criticized by other groups on the grounds that there were too many differences between the induction of mutations by radiation and chemicals to make a comparison simply based on end-point, i.e., the equivalence of the frequency of mutations.

More recently this concept has been reconsidered, particularly by Ehrenberg and his colleagues, as the radequivalence with reference to the genetic or cancer risk from chemical exposures. Again, there has been much discussion and criticism (for a general discussion see "Radiobiological Equivalents of Chemical Pollutants," IAEA Vienna 1980). Since the application of the concept could be of great importance, it should be considered when an analysis of the possible cancer risk to man from chemical exposure is attempted. If it is a valid approach, it would certainly simplify risk analysis.

The radequivalent is the concentration of any particular chemical agent that can induce a mutation frequency or cancer incidence equivalent to 1 rad of acute γ-radiation, or similarly, expressing the mutation or tumor frequency for a dose of the chemical agent in terms of the number of rads of acute γ-radiation giving the same mutation or tumor frequency. This might seem to be a simple way of estimating risk, and, as argued by the proponents, perhaps the only way currently available because of the paucity of information on chemically induced tumors in man. However, the assumptions that have to be made in order to derive the radequivalent values make the approach open to criticism, and indeed of rather little value. It should be added that in the future, when considerably more information is available, such an approach might be valid.

What are the assumptions that have to be made for the estimation of cancer risk to man using radequivalence, and why are these open to criticism? Since the radequival-

Table 10
ESTIMATED PROPORTION OF RESPONDERS IN
TABLES 6 TO 9 (CORRECTED FOR EARLY
DEATHS) AND SE OF THE ESTIMATE[260]

Level	Proportion responders	SE
0 ppm actual in test (0 ppm equivalent lifetime dose)	0.3034	0.0414
10 ppm actual in test (1.79 equivalent lifetime dose)	0.3292	0.0590
33 ppm actual in test (5.89 equivalent lifetime dose)	0.4592	0.0694
100 ppm actual in test (17.86 equivalent lifetime dose)	0.5431	0.0729

ence values are derived from data on mutation induction, the primary assumption when estimating cancer risk is that cancers are initiated by forward mutations in somatic cells. This might be the case, in part, but it is clearly not applicable in all cases. Recent studies have shown that specific chromosome translocations, involving movement of specific oncogenes, are important in blood cancers such as leukemias and lymphomas. There is also evidence of a two-step process of initiation and promotion, that does not allow for simple extrapolation from mutation frequency data. Thus, extrapolation from mutation frequencies to cancer incidence is not legitimate.

The second major assumption is that mutation frequency data obtained for chemical agents can be compared to that for radiation, in order to determine a radequivalence. Such an assumption suggests a similarity in the mechanism of induction of mutations by chemical agents and radiation. However, since the primary lesions induced by the majority of chemical agents are very different from those induced by radiation, and the characteristics of the processes by which the induced lesions are repaired are known to be different, a direct extrapolation is not possible. Most agents that may be expected to effect mutations will do so by an ionic chemical reaction, most often typified by alkylation at an active nucleophilic site, e.g., a DNA molecule. Depending on the target site, some or all of the following events will occur upon exposure to a given chemical: absorption, distribution, and metabolism, resulting in activation or detoxification of the parent molecule. Such a process differs markedly from the kind that occurs as a result of ionizing radiation, in which direct, first order random conversion of chemical species within cells will occur by such chemical processes as coupling (or uncoupling) via free radical reactions. Many chemical agents require metabolic activation to a mutagenic or carcinogenic intermediate, and such activation can be organ or tissue specific — a clear difference of action between radiation and a large subset of chemical agents. Different chemical agents induce different mutation spectra and show different specificities for different cell types. Conversion of mutation frequencies into radequivalent implies a commonality that does not exist.

The third major assumption is that the radiation-induced cancer risk coefficient can be converted into the risk coefficient for chemically induced cancers by multiplication by the radequivalent dose. This again requires, for radiation and chemicals, a similarity of mechanisms of induction, a similarity of dose distribution in the body (unless tissue is specifically measured), a similarity of sensitivity of specific organs and tissues, and a similarity of spectrum of types of tumors. Furthermore, where radiation-induced effects have been shown to be linear to the applied dose, this is certainly not the case for chemically induced alterations. It has not been established that any of these situations exist, and therefore, at this time such an assumption is invalid.

It is premature to consider estimating cancer risk (or genetic risks) to man from

Table 11
PERITONEAL MESOTHELIOMA (PM) IN MALE
RATS: BRRC STUDY

Nominal dose (ppm)	0	10	33	100
Continuous dose (ppm)	0	1.79	5.89	17.86
No. of rats at risk (estimated)	192	96	90	91
No. of responders	4	3	7	22

chemical exposure by utilizing radequivalence. There are too many assumptions and intangibles, such that any risk coefficient obtained has little or no meaning. Some have argued that this represents the only plausible way presently to make cancer risk estimates; but if this way does not have a scientific basis, its plausibility does not make it acceptable.

IV. A BIOLOGICAL PERSPECTIVE ON EO

It is important to recognize that while mathematical dose response models may be useful in the hazard evaluation process, scientists have expressed serious reservation about reliance on numbers generated by models which fail to accommodate the complexities of integrated biologic systems. The purpose of this section is to weigh the biological evidence on EO and to attempt to place it in perspective in relation to other alkylating agents, and in comparison with compounds whose carcinogenic potential for man or animals, or both, is generally regarded as beyond dispute. In considering those biological end-points associated with exposure to EO, which may be useful for purposes of hazard assessment, the following five logically present themselves on the basis of previously reported bioassays: neurotoxicity, teratogenicity, reproductive effects, genotoxicity, and carcinogenicity.

A. Neurotoxicity
Central and peripheral neurological effects of EO are manifested in man and animals exposed to high levels, especially if exposure takes place repeatedly. Neuromuscular effects in mice were elicited at 50 ppm. The highest concentration not producing hindlimb paralysis in several species is ∼100 ppm. For hazard assessment purposes, the rather high dose levels eliciting a response eliminate neurotoxicity as a sensitive indicator of EO toxicity. Epidemiologic studies and other observations of exposed workers, taken together with the results of animal studies, make it likely that no neurological effects occur at or below atmospheric levels of 10 ppm EO.

B. Teratogenicity
Teratogenic effects in rats were not observed via inhalation at the 100 ppm level;[87] dose levels up to 36 mg/kg were not teratogenic in rabbits.[89] Effects observed after administration of 150 mg/kg i.v. in mice[88] are of questionable significance, due to the marked maternal toxicity at this dose level. The very detailed NIOSH study[90] went to great lengths to bring to light any potentially adverse effect that EO might exert on reproduction, embryotoxicity or teratogenicity. Rabbits revealed no such effect at 150 ppm EO, 7 hr/day for various periods before and/or during gestation. Rats manifested severe maternal toxicity, with consequent adverse effects on fetal growth and development, but no teratogenic effects. For quantitative assessment purposes, teratogenicity is not an appropriate end-point, as EO appeared to be teratogenic in one study only, at a level which was markedly toxic to the pregnant females.

C. Effects on Reproduction
Evidence that EO, when absorbed into the body, can reach and penetrate into the

Table 12
PRIMARY BRAIN NEOPLASMS (PBN) IN MALE
RATS: BRRC STUDY[a]

Nominal dose (ppm)	0	10	33	100
Continuous dose (ppm)	0	1.79	5.89	17.86
No. of rats at risk (estimated)	181	92	85	87
No. of responders	1	1	5	7

[a]   Combining the three types of PBN listed in Table 7 (Chapter 8), is
   contrary to the recommendations of the Ad Hoc Panel.[148] However,
   the results for the gliomas would be very similar.

male, and probably the female, reproductive organs makes it imperative to consider
the potential effects of EO exposure on reproduction. Data from a single-generation
reproduction study in rats do indicate an effect at the highest level of exposure, 100
ppm (6 hr/day, 5 days/week for 12 weeks, then 7 days/week for a further 8 weeks).
This level resulted in a decreased number of pups per litter; no effect was seen at 33 or
10 ppm. An increase in fetal malformations was not evident at levels of 100, 33, or 10
ppm, administered by inhalation.

No reproductive effect has been reported in men who were occupationally exposed
to EO. A report of gynecological disorders, spontaneous abortions, and toxemias of
pregnancy in women workers exposed to EO[85] has not been confirmed elsewhere, except for a recent claim that increased rates of spontaneous abortions had been demonstrated in hospital sterilization workers.[86] In both reports insufficient information is
provided to permit interpretation of the results.

One can conclude that no reproductive effects have been demonstrated in the workplace population exposed to EO, and it seems unlikely that EO exposure at or below
10 ppm would give rise to such effects.

D. Genotoxicity of EO

The genotoxic action of EO is expressed as gene mutations in submammalian and
mammalian test systems, and includes the production of dominant lethality and heritable translocations. There is evidence of DNA damage and chromosomal effects elicited by exposure to EO. Observations on SCEs have been discussed.

An issue that arises in attempting hazard assessment is the interpretation of no-
threshold effects seen with in vitro systems. A very different situation exists in vivo,
both with regard to penetration across membrane barriers and the variety of factors
that tend to neutralize the genotoxic action of EO. Unfortunately, the published literature on in vivo systems is confusing because of the variety of routes of administration
and the failure to establish the validity of a claim of cytogenetic effects in male rats
exposed to 2 ppm EO for 66 days.[128] However, with this single exception, it seems
likely that genotoxic actions of EO are expressed in vivo only at atmospheric levels
above 10 ppm, and possibly above 25 ppm.

E. Carcinogenicity of EO

In considering the carcinogenic action of EO, one has to take into account its capacity to bring about tissue injury[151] — for instance under the conditions of repeated
injection into the same subcutaneous tissue site in mice[143] or on repeated gavage of a
solution of EO in oil into the stomach of rats.[144] While tumors were elicited in both of
these studies, some doubt remains concerning the pathogenesis of these neoplasms. If
EO were a gastric carcinogen, its behavior would be expected to resemble that of pro-
totype compounds such as polycyclic aromatic hydrocarbons or N-nitroso compounds

Table 13
MCL IN MALE RATS: NIOSH STUDY

Nominal dose (ppm)	0	50	100
Continuous dose (ppm)	0	10.42	20.80
No. of rats at risk	21	26	15
No. of responders	7	11	10

(for example, N-methyl-N-nitrosourethane) which give rise to tumors within 6 months. In the case of aristolochic acid, malignancy is revealed after 3 months.[228,229] In the experiment by Dunkelberg[144] β-propiolactone (BPL) was used as a positive control and produced tumors after 32 weeks. However, the first tumor with EO was apparent at 79 weeks. In fact, there was a wide divergence between the mortality of BPL rats on the one hand, and the controls and EO rats on the other. Without detailed study of the sequence of events that develop in the course of repeated intubation of rats with EO in oil, it is difficult to decide on the interpretation of gavage studies of this sort.

Of the two completed long-term studies,[27] the BRRC study was conducted in accordance with modern standards, and involved inhalation of EO (6 hr/day, 5 days/week for approximately 2 years). There was an increased incidence of MCL late in the study in male and female rats exposed to 100 or 33 ppm EO, and of peritoneal mesothelioma in male rats at the same levels. No statistically significant effect on these tumors was observed at 10 ppm, even though statistical trend analysis strongly suggests that the leukemic response was also present in females at this level (however, see Appendix C). The incidence of MCL in groups of historic control (untreated) Fischer-344 rats is so high and variable as to encompass all the 10 ppm results seen in this study.

The other completed long-term study, sponsored by NIOSH,[122,122a] yielded results that were essentially similar to those of the BRRC study. These results have been reviewed in Chapter 8 and discussed also in Appendix B.

With respect to the appearance of primary brain neoplasms (Tables 6 and 7, Chapter 8), if the results are compared with the data provided by Koestner,[230] the incidence of tumors in the control groups may be too low, possibly because the search for these lesions did not involve a sufficient number of blocks, or a sufficient number of sections from each block. Applying the criteria put forward by Koestner for a carcinogenic action in the brain, the ability of EO to increase brain tumor incidence reliably and consistently seems to be demonstrated by comparison of the detailed findings in the BRRC and NIOSH studies. As judged by Tables 6 and 7 (see Chapter 8), the BRRC study tumors did not appear earlier in the lives of the experimental animals, as would be expected if EO were acting as a neurocarcinogen. A dose-response relationship is discernible, but assessment of the critical shift to more anaplastic tumor types in the treated groups does not reveal any such effect; a number of other criteria of neurocarcinogenicity are not met by the EO data (see Appendix B). In sum, the issue of an effect of EO on the development of primary brain tumors in the Fischer 344 rat reveals at worst a weak action or, more likely, a modulation of the spontaneous occurrence of gliomas and other central nervous system tumors.

Nevertheless, the effects observed fall within the scope of the accepted definitions of an animal carcinogen. The same augmentation of spontaneous tumor incidence could perhaps have been brought about purely by a promoting effect or other form of modulating action. In a preliminary report on the NIOSH inhalation study, Lynch et al.[122] noted an increased incidence of leukemia in male rats and there was also evidence of an increased incidence of gliomas.[122a] Thus, taking into account the genotoxic properties of EO, it is likely that EO does behave in this instance as an animal carcinogen.

The epidemiological evidence on carcinogenicity of EO in man leaves the issue unsettled. Early reports of cases of leukemia of various kinds, based on observations in

Table 14
RESULTS OF FITTING THE ONE-HIT MODEL TO THE FIVE DATA SETS FOR EO[a]

Data set	1	2	3	4	5
Species	Rat	Rat	Rat	Rat	Rat
Sex	F	M	M	M	M
End-point	MCL	MCL	PM	PBN	MCL
B. E. of α	0.18	0.37	0.02	0.01	0.37
B. E. of β	0.0576	0.0256	0.0131	0.0054	0.0254
p-value for fit	0.99	0.71	0.69	0.55	0.43
SE for β	0.0118	0.00920	0.00293	0.00177	0.0161
Lower 95% CL for β	0.0345	0.0076	0.0074	0.0019	0.0000
Upper 95% CL for β	0.0807	0.0436	0.0188	0.0088	0.0570

Note: Abbreviations: B. E., best estimate; CL, confidence limits.

[a] Data sets 1 to 4 are based on BRRC results; data set 5 is derived from NIOSH results.

small groups of workers exposed to EO as well as other materials, have not been expanded by subsequent cross-sectional, cohort or retrospective studies. Inevitably, each and every study is subject to various criticisms. Despite these shortcomings, the overall picture that emerges suggests that, if EO is carcinogenic in people, its action is probably so weak as to render detection difficult if not impossible in the small cohorts that are available for study.

V. COMPARISON OF EO WITH VINYL CHLORIDE

The metabolic studies described in Chapter 3 revealed the existence of the minor pathway leading to 2-chloroethanol and thence to chloroacetaldehyde, S-(2-hydroxyethyl)cysteine, S-carboxymethylcysteine and thiodiacetic acid. These are the very same metabolites produced by vinyl chloride by way of chloroethylene oxide;[42] together with the corresponding mercapturic acids that appear in the urine. All that is known about the relative proportions of these metabolites formed from EO stems from the work of Jones and Wells.[34] These authors showed that in rats 43% of the radioactivity of ^{14}C-labeled EO (2 mg/kg) was excreted in urine over 50 hr as 9% S-(2-hydroxyethyl)cysteine and 33% as its N-acetyl derivative. When Jones and Edwards[231] administered S-(2-hydroxy[U-^{14}C]ethyl)cysteine to rats (50 mg/kg), the N-acetyl derivative was the major urinary metabolite identified; only 1% of the administered radioactivity was excreted as N-acetyl-S-carboxymethylcysteine. (Note, however, that thiodiacetic acid was not identified.) These results suggest a minor role for the chloroacetaldehyde pathway in EO metabolism, but throw no light on the formation of DNA adducts by this route.

Hathway[232] attributes the mechanism of carcinogenesis and mutagenesis of vinyl chloride to imidazo-cyclization of deoxyadenosine (dA) and deoxycytidine (dC) in rat liver DNA, forming the relatively persistent product etheno-dC and the less persistent etheno-dA. These adducts induce transversions which are consistent with base-pair-substitution (BPS) mutations induced by metabolically activated vinyl chloride, chloroethylene oxide and chloroacetaldehyde. As noted in Chapter 7, EO induces BPS mutations without activation, especially in *Salmonella typhimurium* strain TA 1535.

In spite of these considerations, EO and vinyl chloride have little in common with regard to their toxic effects. The capacity of vinyl chloride to produce hepatic angiosarcomas both in man and animals contrasts with the apparent absence of a target site

Table 15
THE CONTINUOUS LIFETIME EQUIVALENT FOR VARIOUS INDUSTRIAL EXPOSURE SITUATIONS

Actual TWA, exposure[a] (ppm)	Years on the job		
	15	30	45
50	2.29	4.58	6.87
10	0.46	0.92	1.38
5	0.23	0.46	0.69
1	0.046	0.092	0.138
0.5	0.023	0.046	0.069
0.1	0.0046	0.0092	0.0138

[a] While on the job.

in the case of EO. By injection or gavage, EO exercises local effects in whose pathogenesis repeated tissue injury almost certainly plays an essential role. Exposure to EO by chronic inhalation leads to modulation of the natural propensity of Fischer 344 rats to develop spontaneously occurring neoplasia, increasing the incidence of mononuclear cell leukemia, peritoneal mesothelioma and of cerebral gliomas.

The most striking distinction between EO and vinyl chloride is seen in the results of human exposure. Numerous case reports and epidemiologic studies, which have been summarized by IARC,[233] point to liver angiosarcoma as the principal, but by no means the sole or invariable manifestation of the carcinogenic action of vinyl chloride in man. In contrast to these reports and findings with vinyl chloride are the studies on EO reviewed in Chapter 10 and Appendix A. Reports of leukemia and gastric cancer attributed to EO exposure[74,181] have not been borne out by other studies.[75,77] As discussed earlier in this chapter, one cannot conclude that EO has been shown to be a human carcinogen.

What are the general implications of the observations discussed above? The metabolic data on EO suggest the possibility of nonlinear kinetics, in much the same way that large doses of vinyl chloride were shown to saturate the metabolic and transport processes of the body,[234] generating a ''hockey-stick'' type of dose-response curve.[235,236] The implications of nonlinear kinetics with respect to risk estimation have recently been discussed by Hoel et al.[237] These authors illustrate the fact that, in the presence of nonlinear kinetics, the mathematical models conventionally used for low-dose extrapolation are potentially capable of overestimating risk by several orders of magnitude.

VI. HAZARD ASSESSMENT

The situation with regard to EO is fraught with uncertainties which render hazard assessment to a large extent judgmental. It is perhaps preferable to admit this inescapable necessity, rather than appear to rely on mathematical extrapolations which — in the present state of our knowledge of EO molecular biology and pharmacokinetics — cannot be soundly based.

Conclusions are based on the qualitative and quantitative aspects of the various biological phenomena considered. For purposes of worker protection, three zones may be postulated, in terms of uniform, 8-hr/day, 5-day/week, exposure conditions (or equivalent work periods) calculated as time weighted averages. The zones are constructed on the basis of differing degrees of confidence with respect to potential hazard.

Table 16

BEST ESTIMATES OF THE EXCESS CANCER RISK
(% *p*) FROM THE FIVE DATA SETS (TABLES 1 TO 13)
FOR VARIOUS CONTINUOUS LIFETIME
EXPOSURES

Continuous[a] lifetime dose (ppm)	Experimental data set[b]				
	1	2	3	4	5[c]
5.0	2.5–1[d]	1.2–1	6.3–2	2.6–2	1.2–1
1.0	5.6–2	2.5–2	1.3–2	5.3–3	2.5–2
0.5	2.8–2	1.3–2	6.5–3	2.7–3	1.3–2
0.1	5.7–3	2.6–3	1.3–3	5.4–4	2.5–3
0.05	2.9–3	1.3–3	6.5–4	2.7–4	1.3–3
0.01	5.8–4	2.6–4	1.3–4	5.4–5	2.5–4
0.005	2.9–4	1.3–4	6.5–5	2.7–5	1.3–4
0.001	5.8–5	2.6–5	1.3–5	5.4–6	2.5–5
0.0005	2.9–5	1.3–5	6.5–6	2.7–6	1.3–5
0.0001	5.8–6	2.6–6	1.3–6	5.4–7	2.5–6

[a] The "continuous lifetime dose" as obtained from "translating" the actual 8-hr. (or other) dose to the continuous lifetime equivalent, as from Table 15.

[b] Data sets 1 to 4 are based on BRRC results; data set 5 is derived from NIOSH results.

[c] In this use of the model, it became apparent that the data on the male rats from the NIOSH study gave essentially the same number for predicted excess risk based on MCL data as did the results from the BRRC study; consequently, for purposes of further discussion, the NIOSH male MCL data are not itemized separately but may be inferred to duplicate the BRRC male MCL data.

[d] "2.5 –1" is 2.5×10^{-1} or 2.5/10; "6.5 –4" is 6.5×10^{-4} or $6.5/10^4$

A. Zone of Increased Probability for Potential Adverse Health Effects in Man

As indicated, observation of SCEs in peripheral lymphocytes of exposed personnel is a convenient monitor of exposure, but the health significance of an increase in the level of SCEs is unclear. Taking other possible effects into consideration (genotoxicity, carcinogenicity, reproduction, neurotoxicity), it is highly probable that these would only occur — if at all — at levels above 10 ppm. It should be stressed that no adverse health effect has been reliably demonstrated in man following exposure to 10 ppm EO.

B. Zone of Uncertain Consequences

This is essentially a zone of ignorance because of lack of data, the existence of strikingly nonuniform exposures in some instances, and a variety of other known and unknown factors. This zone is considered to exist for TWAs within the range of 1 and 10 ppm. Conceivably the zone could extend from 3 or even 5 ppm to 10 ppm. At present it is impossible to set the boundaries more precisely.

C. Zone of Inconsequential Exposure

From the standpoint of potential health effects in man, exposure at or below 1 ppm TWA may be considered as presenting no apparent hazard. An increase in this exposure limit may be possible when the acquisition of additional scientific knowledge concerning EO justifies such a course.

Table 17
PREDICTED EXCESS CANCER RISK FROM EXPERIMENTAL ENDPOINTS

Continuous lifetime dose (ppm)[a]	MCL female rats, BRRC	MCL male rats, BRRC	PM male rats, BRRC	PBN male rats, BRRC
0.046	2.6×10^{-3}	1.2×10^{-3}	6.0×10^{-4}	2.5×10^{-4}
0.092	5.3×10^{-3}	2.4×10^{-3}	1.2×10^{-3}	5.0×10^{-4}
0.138	7.9×10^{-3}	3.5×10^{-3}	1.8×10^{-3}	7.4×10^{-4}

[a] Continuous lifetime dose calculated for a worker exposed to EO by inhalation over 15, 30, or 45 years to 1.0 ppm TWA_s, 5 day/week, 48 week/year.

$$
\begin{aligned}
\text{Continuous lifetime dose} &= 1 \times 15/72 \times 8/24 \times 5/7 \times 48/52 = 0.046 \\
&= 1 \times 30/72 \times 8/24 \times 5/7 \times 48/52 = 0.092 \\
&= 1 \times 45/72 \times 8/24 \times 5/7 \times 48/52 = 0.138
\end{aligned}
$$

Table 18
COMPARISON OF PREDICTED EXCESS RISK FROM OSHA AND EOIC TREATMENTS OF EXPERIMENTAL ENDPOINTS

Actual exposure level (ppm)	OSHA treatment[a]			EOIC treatment[b]	
	MLE[c] multi-stage male PM	MLE one-hit male PM	MLE multi-stage female MCL	One-hit male PM	One-hit female MCL
5.0	5.8×10^{-3}	7.7×10^{-3}	1.15×10^{-2}	9.0×10^{-3}	3.9×10^{-2}
1.0	1.2×10^{-3}	1.5×10^{-3}	2.3×10^{-3}	1.8×10^{-3}	7.9×10^{-3}
0.5	6.0×10^{-4}	8×10^{-4}	1.2×10^{-3}	9.0×10^{-4}	3.9×10^{-3}

[a] Excess risk, lifetime exposure assumed to be 8 hr/day, 5 day/week, 46 week/year for 45 years in a 54-year lifespan since initial exposure.

[b] Excess risk, 8 hr/day, 5 day/week, 48 week/year, for 45 years of an expected 72-year lifespan. The "Actual Exposure Level (ppm)" is for the workplace atmosphere under the foregoing conditions, and was translated for purposes of the calculation to the appropriate "Continuous Lifetime Equivalent Dose".

[c] MLE is "Maximum Likelihood Estimate of Excess Risk"; multistage refers to the multistage model, one-hit to the one-hit model; "male PM" is based on male peritoneal mesothelioma data from the BRRC study, and "female MCL" is based on female mononuclear cell leukemia data from the BRRC study.

Table 19
ESTIMATED VALUE OF THE SHAPE PARAMETER (k) FOR THE TIME TO A TUMOR (MCL) FOR EACH OF THE DOSE GROUPS IN THE BRRC ANIMAL STUDY

Sex	Nominal dose (ppm)	Tumor risk at 18 months	Tumor risk at 25 months	Estimated value of k
F	0	0	0.1704	Infinity
	10	0	0.2438	Infinity
	33	0.0111	0.4053	11.7
	100	0.05	0.7074	9.7

Table 19 (continued)
ESTIMATED VALUE OF THE SHAPE PARAMETER (k) FOR THE TIME TO A TUMOR (MCL) FOR EACH OF THE DOSE GROUPS IN THE BRRC ANIMAL STUDY

Sex	Nominal dose (ppm)	Tumor risk at 18 months	Tumor risk at 25 months	Estimated value of k
M	0	0.0108	0.3035	10.6
	10	0	0.3292	Infinity
	33	0.0325	0.4592	8.9
	100	0	0.5431	Infinity

Note: Sample calculation for the females at 100 ppm:

$$k = \frac{\ln(-\ln(1 - 0.7074)) - \ln(-\ln(1 - 0.05))}{\ln 25 - \ln 18} = 9.7.$$

Median tumor-free lifespan (MTFL) = $0.5^{1/k}$. If for example, $k = 10$, then MTFL = $0.5^{1/10}$ = 0.933. This implies that the typical animal that developed a tumor did so at 93% of its nominal lifetime (25 months for the rats in this experiment). If only 1% of the animals at risk actually developed a tumor, then the median tumor-free lifespan for the group as a whole is $0.01 \times 0.933 + 0.99 \times 1.0 = 0.99933$. Thus, the average member of the group loses 0.00067 (or 0.067%) of its tumor-free lifetime.

Table 20
CALCULATED EXCESS RISKS (% p) FOR THE AVERAGE SWEDISH WORKER'S EXPOSURE BASED ON THE RESULTS FROM THE FIVE ANIMAL END-POINTS

Data set	1	2	3	4	5
Species	rat	rat	rat	rat	rat
Sex	F	M	M	M	M
End-point	MCL	MCL	PM	PBN	MCL
B. E. of β	0.0576	0.0256	0.0131	0.0054	0.0254
Lower 95% CL for β	0.0345	0.0076	0.0074	0.0019	0.0000
Upper 95% CL for β	0.0807	0.0436	0.0188	0.0088	0.0570
Using k = 10					
B. E. of %p	0.016	0.0071	0.0037	0.0015	0.0071
Lower CL for %p	0.010	0.0021	0.0021	0.0005	0.0000
Upper CL for %p	0.022	0.012	0.0052	0.0025	0.016
Using k = 6.5					
B. E. of %p	0.027	0.012	0.0061	0.0025	0.012
Lower CL for %p	0.016	0.0036	0.0035	0.0009	0.0000
Upper CL for %p	0.037	0.020	0.0088	0.0041	0.026

Notes: Data sets 1 to 4 are based on BRRC results; data set 5 is derived from NIOSH results. Abbreviations: B. E., best estimate; CL, confidence limits.

Table 21
BEST ESTIMATE OF THE SHAPE PARAMETER (m) IN THE WEIBULL MODEL AND THE ONE-SIDED p-VALUE FOR EACH TREATED GROUP FOR EACH OF THE FIVE ANIMAL END-POINTS

Data set 1 (m = 1.04)				
CLD	0	1.79	5.89	17.86
PR	0.1704	0.2438	0.4053	0.7074
SE	0.0332	0.0568	0.0644	0.0741
One-sided p-value		0.13	0.0006	0.0001
Data set 2 (m = 0.76)				
CLD	0	1.79	5.89	17.86
PR	0.3034	0.3292	0.4592	0.5431
SE	0.0414	0.0590	0.0694	0.0729
One-sided p-value		0.36	0.027	0.0062
Data set 3 (m = 1.33)				
CLD	0	1.79	5.89	17.86
PR	0.0208	0.0313	0.0778	0.2418
One-sided p-value		0.45	0.024	0.0001
Data set 4 (m = 0.77)				
CLD	0	1.79	5.89	17.86
PR	0.0055	0.0109	0.0588	0.0805
One-sided p-value		0.56	0.014	0.0018
Data set 5 (m = 2.26)				
CLD	0	10.42	20.83	
PR	0.3333	0.4231	0.6667	
One-sided p-value		0.37	0.051	

Note: Data sets 1 to 4 are based on BRRC results; data set 5 is derived from NIOSH results. Abbreviations: CLD, continuous lifetime dose (ppm); PR, proportion of responders.

Appendix A

LITERATURE REVIEW OF HUMAN HEALTH EFFECTS OF ETHYLENE OXIDE EXPOSURE

I. CHRONIC TOXICITY OF EO"

The purpose of this study was to evaluate the chronic health effects of long-term, low-level exposure to EO in a group of chemical operators engaged in the manufacture of EO at a Texas chemical plant. The production process involved direct oxidation of ethylene gas to oxide. The study group consisted of 37 of 40 EO operators at the plant in 1963. These operators were aged 29 to 56 years, were hired between 1941 and 1962, and their length of employment in the EO units ranged from less than 5 years to 16 years. There were eight former operators who worked a minimum of 100 months in the EO unit also included. The control group consisted of 40 other operators at the plant who had never worked in the EO unit. These were similarly aged and worked an average of 11 2/3 years at the plant. EO was extensively sampled in 1962 to 1963 by the Bolton and Ketcham impinger collection technique and operator exposure was estimated at 5 to 10 ppm (TWA$_8$).

Potential health effects of EO exposure were investigated through physical and laboratory examinations, occcupational medical record reviews, and reviews of medical absenteeism experience. Laboratory studies included red blood cell (RBC) and white blood cell (WBC) counts, hemoglobin (Hb), routine urinalysis, urine bilirubin, cephalin flocculation, and icteric index. Former EO operators were not given a physical examination.

No differences in health status were observed between EO and non-EO operators based on physical examination. Similar results were obtained for EO and non-EO operators for mean RBC (5.21 vs. 5.20 million), mean WBC (9.124 vs. 7.553 thousand) and mean Hb (15.4 vs. 15.3) values. Of the 40 EO operators one had RBC less than 4.0 million and a Hb level less than 13.6 grams. All RBC and Hb values for non-EO operators were within the normal range. There was no indication of kidney or liver abnormalities among EO operators based on investigations for albuminuria, cephalin flocculation tests, and icteric index.

A review of medical histories for the period 1955 to 1962 revealed no significant differences between EO and non-EO operators. Although slight anemia was diagnosed for three EO operators during this period, a similar diagnosis was made for six control operators. Among EO operators, four neoplasms occurred: three benign neoplasms and one adenocarcinoma of the bladder. Among control operators there were six neoplasms: two basal cell skin cancer, one testicular seminoma, and three benign neoplasms. No evidence of neurological disorders among EO operators was found; diagnoses of "mental, psychoneurotic and personality disorders" were documented in medical records of five EO operators and four non-EO operators.

A. Critique

This study suggests that EO exposure at the levels measured in this study has no noticeable effect on the hemopoietic system. It also suggests that low-level exposures may not be associated with acute clinical disorders (especially those affecting the liver or kidneys) that may be expected to occur after a relatively short latent-induction period.

This study is of little use, however, in ruling out the possibility that low-level exposure to EO may have long-term chronic health effects. Chemically induced cancer, for example, may have a latent-induction period of 10 to 40 years, whereas some workers in this study were observed for only 5 years. This limitation and the small number of employees examined suggest that the results of the study be interpreted cautiously as far as chronic diseases are concerned.

II. HEMATOLOGIC STUDIES ON PERSONS OCCUPATIONALLY EXPOSED TO EO[183]

A. Hematologic Investigation

This is a report of results of a 1961 study of workers in a Swedish factory which manufactured and used EO. It was conducted because an elevation in WBC counts was found in a 1960 survey among some of these same workers who were exposed to EO. In this study, hematologic examinations were performed on four groups of workers classified as "exposed" (n = 37), "previously exposed" (n = 54), "intermittently exposed" (n = 86), and "nonexposed" (n = 66). "Exposed" persons were from part of the factory where leakage of EO from tube joints and pumps occasionally occurred. This group's exposure ranged from 2 to 20 years and included a small number of persons who had been accidentally exposed to high concentrations of EO during their work. The control group consisted of similarly aged persons from other* departments. "Healthy" persons (defined as those with observed sedimentation rates of less than 12 mm/hr) were analyzed separately from those considered "unhealthy".

Comparison of hematologic findings revealed significantly lower percent Hb values among exposed (92.1) vs. control (95.2) workers. Although not significant, there were three cases of anisocytosis among exposed workers (which the authors interpret as suggesting the possibility of depressed bone marrow function). Among the exposed group there was one case of lymphatic leukemia (? chronic). A significant elevation in WBC counts for EO exposed workers was also found, compared with controls, with over 50% of the exposed group having higher WBC counts than controls. However, all counts for exposed workers were within the normal range. Higher lymphocyte counts among exposed as compared to control individuals were also noted, but the difference was not statistically significant. Among "healthy" persons, lymphocyte counts (per mm^3) were 2192 for exposed, 2126 and 2036 for intermittently and previously exposed, and 2016 for the nonexposed group. The authors interpret these results as a dose-response gradient suggesting that some type of lymphocyte stimulation occurred, which may be attributable to EO exposure. They also conclude that some reversibility may have occurred between 1960 and 1961 due to improved ventilation and safety controls. This conclusion was based on the reduction of lymphocyte counts in 13 individuals from a mean of 2700 in 1960 to a mean of 2210 in 1961.

1. Critique

Although the authors suggest that the lymphocyte results indicate a dose-response relationship, such an inference is difficult to defend medically, when one considers the small number of individuals examined, the small differences detected, and the fact that all observed data points fell within the normal range. Even if the lack of clinical significance is ignored, it is not possible to infer from this study the levels at which such changes might be induced. It is likely that exposure levels of employees in the exposed group ranged widely, since some employees had worked for as little as 2 years at low-level exposures and others had extremely high exposures as a result of accidental contamination. Although the authors note the reduction in mean lymphocyte counts that

---

* A reference on page 329 states that controls were from "other" departments working with EO. No clarification is provided regarding why these persons were considered to be nonexposed.

occurred among exposed individuals between 1960 and 1961, it is curious that they do not note the reductions (1980 to 1730) which also occurred in nonexposed employees between 1960 and 1961. Neither can the reduction in Hb levels be taken seriously since no dose-response relationship is evidenced with only two groups, and the reported difference between exposed and nonexposed workers lacks clinical significance.

There are three other problems which should be noted. The first involves the lack of information to verify that exposed and nonexposed groups were representative of the overall plant populations in each category. This requires additional information on the manner in which the groups were selected and how eligible employees were identified. A second problem is that insufficient information is provided with which to judge the appropriateness of the statistical analyses performed. A third problem is the concern that control employees may have also been exposed to EO since 26 persons from other departments working with EO were included in the control group.

B. Chromosomal Aberrations

The potential of EO exposure to induce chromosomal aberrations in somatic cells was evaluated in seven out of eight persons who had received high-level exposures to EO over a period of 2 hr during an accident in a Swedish factory. "Gross" (referring to structural and numerical aberrations excluding gaps) and "pathological" aberrations (including gaps) were compared in phytohaemagglutinin-stimulated lymphocytes from the seven exposed and ten nonexposed (control) persons from the same factory. For exposed subjects, blood samples were taken 18 months after the acute exposure occurred. The number of metaphase plates analyzed per person ranged from 6 to 26.

The results indicated that the exposed individuals had higher rates of chromosomal aberrations than control individuals. The number of pathological aberrations per 100 cells for exposed individuals was 30.2 compared with a rate of 16.5 for nonexposed individuals. Similarly, the rate of gross aberrations per 100 cells was 17.5 for exposed and 4.3 for unexposed individuals.

1. Critique

Although there appear to be significant differences in chromosomal aberration frequencies between highly exposed and nonexposed control workers, several limitations in the data as presented should be recognized. First, since all tests were performed only after exposures occurred, there is a possibility that elevated aberration frequencies existed before exposure as well. No information is provided regarding simultaneous exposures to other agents which may have contributed to the elevation in aberration frequencies in these individuals. Second, rates based on inadequate numbers of cells in only two groups are not sufficient to demonstrate conclusively a dose-response relationship. Third, exposed and control workers may have been evaluated at different times, a procedure which could have introduced some artifactual variation into the data set.

III. GYNECOLOGICAL SICK RATE OF WORKING WOMEN OCCUPIED IN THE PRODUCTION OF EO[85]

The purpose of this study was to investigate the effect of exposure on the "gynecological sick rate" and "course of pregnancy" in female factory workers. The setting for this study was a Russian polyethylene production unit involved in catalytic synthesis of EO. It is stated that EO exposures never exceeded 0.5 ppm. The effects attributable to EO in this study are

1.   An increased "gynecological sick rate"

2.   "Unfavorable effects on female reproductive organs" (including "erosion of uterine cervix" and "inflammatory diseases of the uterus and hypophysis")
3.   Effects on pregnancy (including "pregnancy interruption" and "toxemia of pregnancy")

A. Critique

It is unclear from translations of this report how the study was conducted and what methods were used to ascertain and categorize health effects of study participants. There are other problems, as well, which make the results difficult to interpret. For example, the medical terms used lack sufficient clarity to know what diseases and disorders were observed. Numbers and classifications of exposed and nonexposed groups lack consistency throughout the text. No information is given regarding when this study was done and over what period of time. These problems preclude any meaningful assessment of this report.

IV. LEUKEMIA IN WORKERS EXPOSED TO EO[181]

This is a report of the occurrence of leukemia among workers in a Swedish factory which used a mixture of 50% methyl formate and 50% EO to sterilize hospital equipment. The EO mixture was first used at this factory in 1968. Although few persons worked directly with the sterilization mixture, others were exposed because of off gassing from boxes containing the sterilized equipment. In 1978, approximately 30 women worked full-time in a hall where the boxes were stored; another 100 had occasional exposure from passing through the hall. Between 1968 and 1977, 70 persons were reported to have worked for some period of time in the storage hall and another 160 in neighboring rooms or in the sterilization room. These 230 workers formed the study cohort. Air sample monitoring conducted in 1977 indicated that short-term exposures in the storage hall ranged from 2 to 70 ppm with an estimated TWA_8 of 20 ± 10 ppm. Reported short-term EO levels inside newly sterilized boxes were as high as 1500 ppm and outside floor samples were as high as 150 ppm. Exposure in the storage hall was reported to be higher than that in the sterilizer room.

A total of three cases of leukemia were diagnosed among plant employees between 1972 and 1977, 4 to 9 years after the introduction of EO in the sterilization process. Case 1 began work in the storage hall less than 5 years prior to a diagnosis of chronic myeloid leukemia. Case 2 was diagnosed as acute myelogenous leukemia approximately 9 years after first working in the storage hall. Case 3 was diagnosed with "macroglobulinemia" approximately 6 years after first exposure. In contrast to Cases 1 and 2, who were women, Case 3 was a male plant manager whose exposure to the EO mixture was estimated at 3 hr/week with occasional exposure to benzene in laboratory work. Among the estimated 230 persons who worked in exposed areas, an estimated 0.2 cases of leukemia would have been expected using Swedish national leukemia incidence rates for 1972. The author concluded from these observations that EO is a leukemogen capable of inducing a broad spectrum of cytogenetic aberrations in proliferating hemopoietic cells.

A. Critique

Caution must be exercised in interpreting the expected number of leukemia cases reported in this paper as these were based on an "estimated" number of exposed workers — not on complete ascertainment of persons and person/years in a well-defined cohort. There was no attempt to validate the number of potentially exposed personnel, thus the 230 persons may be an inaccurate estimate. Secondly, expected incidence was calculated only for employees working in the storage hall and adjacent areas, whereas one of the cases (Case 3) was a plant manager assigned to neither of these areas. Third, no information is given regarding the length of exposure to EO for Cases 1 and 2 and whether they could have worked opening newly sterilized boxes where exposure could have ranged to 1500 ppm. There is also the problem of confounding exposures to benzene and methyl formate. Because of the small number of cases involved, the methodological problems mentioned, and the likelihood that this investigation was conducted subsequent to identification of the three leukemia cases, it is inappropriate to interpret these results as indicative of the carcinogenic potential of EO. Rather, this case report serves as an indication that further objective studies are needed.

V. A COHORT STUDY OF MORTALITY AND CANCER INCIDENCE IN EO PRODUCTION WORKERS[74]

This is a report of a cohort mortality and cancer incidence study of 241 production plant workers in a Swedish chemical factory which manufactured EO by the chlorohydrin process since 1940. Between 1940 and 1947, workers in the production room were exposed to EO, ethylene, DDT, ethylene glycol and a variety of chlorinated products including ethylene chlorohydrin, ethylene dichloride, and *bis*-chloroethyl ether. Average EO exposures were estimated to have been below 25 mg/m³ with occasional peaks up to the odor threshold (1300 mg/m³). After 1950, exposure increased with more production of EO-based compounds, until 1963 when EO production (but not use) ceased.

This study was conducted as a follow-up to a 1961 study which suggested that EO may cause decreased hemoglobin concentration in RBC and significant lymphocytosis in exposed workers. The workers were divided into three groups: 89 full-time exposed men who had worked directly in the EO manufacturing area; 86 intermittently exposed men assigned to maintenance work; and 66 nonexposed men working in other parts of the factory. The study cohort was limited to workers included in the 1961 study who were employed for more than 1 year at the plant between 1940 and 1977. Their mortality and cancer incidence experience were investigated for the 17-year period, 1961 to 1977. For the exposed workers, follow-up was initiated in 1961 or 10 years following their first EO exposure (if later than January 1961) to allow for a possible 10 year induction-latent period for EO carcinogenic activity.

Overall cancer mortality in the full-time exposed cohort was approximately 2.6 times expected based on 9 observed (O) and 3.4 (E) deaths. No excess was found among intermittently exposed or nonexposed workers. The excess in the direct exposure group was attributable primarily to cancers of the stomach (O = 3, E = 0.4) and leukemia (O = 2, E = 0.14). These employees also had significantly elevated mortality from all circulatory diseases (O = 12, E = 6.3) which was not considered to be exposure related, as a similar excess was seen in the control group. An excess cancer incidence of 1.86 was found based on 11 observed and 5.9 expected cancers in the full-time exposure cohort. The authors interpret these results as suggesting that EO and ethylene dichloride exposure, together with ethylene chlorohydrin or ethylene, may be related to the observed excess cancer at this factory.

A. Critique

As the authors acknowledge, it would be difficult to link EO directly to the alleged increase in cancer incidence reported at this factory because of the multiple exposures of these workers and the lack of good industrial hygiene monitoring data or exposure estimates for the workers involved. In addition, the small number of reported cases requires that caution be exercised in interpreting the results of this study. No 10-year latency period was considered in the analyses of control (or nonexposed) workers; therefore, there is concern regarding the comparability of the results among cohorts.

VI. EO: EVIDENCE OF HUMAN CHROMOSOMAL EFFECTS[73]

The purpose of this study was to assess the mutagenic potential of EO. This was accomplished by analyzing SCE rates in circulating lymphocytes of three groups of workers exposed in a hospital setting during a 5-month period, January to May, 1978. Group I (exposed) included 12 workers — 8 women and 4 men — from the sterilization area; Group II (intermittently exposed) consisted of 12 workers from the adjacent operating room; and Group III (control) was selected from unexposed laboratory staff. Participation in the study was voluntary. Of the 12 workers from Group I, 4 were classified as "symptomatic" because they became ill during the 5-month study period. They were removed from exposure and SCE measurements were made at 4 and 8 weeks after their last known exposure to EO. The remaining 8 workers from Group I had SCE measurements taken between the 8th and 9th week after last known exposure. Group III served as an internal control group; several of these workers had SCEs remeasured during the course of the study to assess individual variation. For each individual, 20 mitoses were scored. Although no personal time weighted EO exposure measurements were made, the maximum ambient EO levels were reported as 36 ppm with peaks of over 1500 ppm in certain areas of the sterilization unit.

From Group I, four individuals reported symptoms of central nervous system effects and upper respiratory irritation. These symptoms disappeared after removal from exposure areas. No similar complaints were reported among controls during the same 5-month study period. Average SCE rates (SCEs per cell) among these four individuals at 3 (9.79) and 8 weeks (10.3) after last exposure were significantly higher than the rate among eight controls (6.37). Rates among the eight nonsymptomatic workers from Group I at 8 to 9 weeks after last exposure were found to be significantly higher (8.65) than controls. From the intermittently exposed group, four individuals also had an increased mean SCE rate (9.37) between 7 and 9 weeks after exposure. The author interprets these results as indicating an increased rate of mutagenic effects in human somatic cells associated with EO exposure and that these effects may persist over time.

A. Critique

Incomplete and unclear data presentation renders the results of this study difficult to interpret. Data presented in the tables refer to selected subsets of Groups I (4 of 12 workers), II (4 of 12 workers), Control Group (8 of 12 workers) and not to the complete study groups. It is, therefore, unclear whether all data were analyzed and, if so, why they were omitted from the tables. Although several controls were remeasured during the course of the study to assess baseline variation in SCE rates, these data are neither discussed nor presented in the paper.

With respect to analyses, it is of interest to know the mean, which was reported, as well as the median (not reported) number of SCEs per cell for each group in order to determine whether these results were attributable to only a few individuals rather than the entire group.

Other problems with the study include the nonuniform manner and timing of SCE testing, and inadequate control for other variables (i.e., smoking, drugs) that could have contributed to increased SCEs. Also, as there was not mention of "blind" SCE readings, the study may have been subject to some degree of "observer bias."

There is reason to discount the significance of these results, because the number of subjects, as well as number of cells examined per subject (6 to 20), were small, reported differences were not large, and exposure levels for subjects are unknown.

VII. STUDIES ON WORKERS EXPOSED TO EO

- Recent studies with workers exposed to EO[184]
- An evaluation of possible effects on health following exposure to EO[169,169a]

This study investigated effects of EO exposure among 75 employees of nine American Hospital Supply Corporation (AHSC) facilities that used EO as a sterilant gas. Reported employee exposure at these facilities was within the OSHA 50 ppm TWA$_8$ standard; short-term exposures occasionally exceeded 75 ppm. The employees studied held a variety of jobs (operators, maintenance mechanics, etc.) involving varying degrees of exposure. Investigations included a complete physical and biological (laboratory) examination, sperm studies, and cytogenetic testing (chromosomal aberrations and SCEs). Exposed workers were subjected to a second examination, performed at a different laboratory, within 12 months from the initial examination. An internal control group was selected from one plant. These included 26 unexposed employees and 15 other employees for whom preplacement biological specimens had been obtained.

Abrahams[184] found no correlation between EO exposure and results of hematologic examinations, urinalyses or other laboratory studies. Although a higher proportion of the exposed workers than controls tested has sperm counts less than 20 million/ml, an insufficient number of exposed (46) and control (9) specimens were available on which firm conclusions could be based. With respect to cytogenetic studies, an increased rate of chromosomal abnormalities was reported for exposed workers vs. controls at eight of the nine facilities. Data for the ninth facility was not reported because the number of workers studied was small and the sterilization procedure used was not comparable with that used in the other eight facilites. "Total aberrations" were elevated as were "unstable forms" which ranged from 1.14 to 1.83/100 cells (initial testing) in exposed workers compared with 0.49/100 cells in control employees. During subsequent testing in 1979, values for unstable forms ranged from to 0.8 to 3.0 among workers at two of the nine facilities.

Mean SCE rates were elevated in workers at five of the nine facilities. The mean value for the 41 controls was 5.35, as compared with 8.04 to 9.15 for exposed workers. A significant correlation between rates of SCEs and quadriradials was also reported. Abrahams interpreted these results as suggesting that certain types of chromosomal aberrations found in these workers might be attributable to EO exposure.

A. Critique

Regarding the results of the physical and laboratory examinations, no data were presented. There is no discussion of whether, or how, a dose-response relationship was investigated or if a control group was used in the analysis. With respect to the results of the cytogenetic tests, irregularities in the design and conduct of the study and apparent incomplete analysis of the data render the evidence presented "suggestive", at best.

Although the data presented in the tables do suggest a significantly higher rate of aberrations in exposed vs. control individuals, no dose-response relationship was investigated. It is therefore uncertain whether this difference can be attributed to EO. Second, incomplete reporting of data causes interpretive problems. For example, overall results for the 75 exposed employees are not presented, nor are data for one of the nine facilities. SCE results are omitted for four plants and results of the second, or "subsequent", testing are omitted for four of the nine plants. Were these plants not part of the retests? A third problem arises with respect to the manner in which the "subsequent" tests were conducted. Although the author stated that the retests were intended to provide some measure of validity for the initial results, he indicated that both the laboratory and testing procedures used for scoring the chromosomal abnormalities differed from the initial tests. As a result, it is possible that some of the changes detected between initial and subsequent tests reflect differences in testing procedures. If the retests were intended to determine the persistence of chromosomal abnormalities over time (and this is not stated), some explanation is needed for not subjecting the controls to subsequent testing as well, so that some measure of the internal consistency of SCEs in nonexposed employees could be obtained. A fourth problem regards the adequacy with which the comparison of group "means" addresses the issue of an exposure-related effect. Because SCEs are not normally distributed, "median" values and/or proportionate distributions of cells with varying numbers of aberrations are more appropriate statistics for comparison. Because of these concerns, sufficient evidence is lacking from this study to conclude that an exposure-related effect was found or to estimate the exposure level which could have caused such an effect.

B. Follow-Up Reports

Richmond et al.[169,169a] have provided updated accounts of the same study that cover the data for 1979 and 1980. Significant differences were found in the numbers and types of chromosomal aberrations between the exposed workers and the nonexposed controls, especially the presence of increased numbers of quadriradial and triradial forms in exposed individuals. Some, but not all, exposed workers had increased numbers of SCEs. Follow-up, over a period of 4 years, of 13 workers removed from exposure because of increased numbers of aberrant cells has not shown distinct improvement in the incidence of chromosomal aberrations, but there was a moderate lowering of the numbers of SCEs. The authors feel that SCEs are a sensitive indicator of recent exposure to EO, in that they are increased in number before chromosomal aberrations appear, and — should exposure cease before chromosomal aberrations become apparent — the numbers of SCEs subsequently decline.

VIII. IN VIVO AND IN VITRO EO EXPOSURE OF HUMAN LYMPHOCYTES ASSESSED BY CHEMICAL STIMULATION OF UNSCHEDULED DNA SYNTHESIS[116]

This study investigated cytogenetic effects of EO exposure among 17 exposed and 11 control workers (all female) in a Swedish factory that sterilized disposable medical equipment. The mixture used to sterilize equipment consisted of 50% EO and 50% methyl formate. The 17 exposed workers included 12 packers, 4 sterilizers and 1 laboratory technician. The 12 packers were exposed repeatedly to bursts of 5 to 10 ppm EO for approximately 1 hr each day. The 4 sterilizers and 1 laboratory technician were exposed to 0.5 to 1 ppm (TWA). The controls (n = 11) were unexposed assembly line workers. Similar proportions of cases and controls were smokers (53 to 54%). The mutagenic potential of EO was investigated by measuring unscheduled DNA synthesis (UDS) in peripheral blood (leukocytes and lymphocytes), where UDS was stimulated by the carcinogen NA-AAF. The frequency of chromosome aberrations in peripheral lymphocytes was also investigated. For this analysis, 200 cells per individual were scored. Individuals were re-examined after improvements in factory ventilation were made which reduced EO concentration to trace amounts.

The authors report that (1) total chromatid gaps plus breaks were significantly elevated and (2) NA-AAF induced UDS was significantly reduced in the EO-exposed groups as compared with the unexposed control group. A negative correlation between duration of EO exposure and the two effects investigated was also reported. This is interpreted as lending support to the hypothesis that EO in vivo exposure inhibits DNA-repair capacity. Following improvements in factory ventilation, induced UDS values returned to normal in both packers and sterilizers but no reduction in chromosome aberrations was detected. The authors suggest that decreases in UDS are attributable to EO cytotoxicity and to decreased response potential in surviving cells.

A. Critique

The results of the studies conducted are poorly described, raising questions regarding their interpretation. With respect to the chromosome aberration data, the authors report "significant" statistical differences but do not address their "clinical" significance. The inclusion of "gaps" in the calculation of mean number of aberrations per 200 cells is questionable in view of the uncertain importance of these to any biological end-points. When gaps are excluded, the mean numbers of aberrations per 100 cells among the exposed and control groups are quite similar (packers = 6.0; sterilizers = 7.8; controls = 5.3).

The differences shown with respect to NA-AAF-induced UDS are of questionable significance. The range of measurements for the two exposure groups of 12 packers and 5 sterilizers are 50 to 319 and 191 to 354 cpm, respectively. These ranges overlap considerably with that of the control group (124 to 500). Given the wide range of individual variability of this test result and the small number of individuals in each group, the reported differences are of questionable significance. Analyses of the data by calculation of means and standard errors is of limited value because chromosome aberrations are not normally distributed. The same criticism applies to the analysis of the association between decreased UDS and increasing length of exposure. The variation among individuals with respect to this variable was not discussed. The in vitro investigation of the effect of EO in UDS was based on only one individual.

Differences in aberration frequencies and UDS values cannot be unequivocally attributed to EO exposure because this is a cross-sectional study which looked at workers at only one point in time and the workers studied were also exposed to methyl formate. Although some comparability of smoking behavior among groups was reported, other factors such as medications and prior infections were not considered as possible causes of cytogenetic abnormalities. Regarding the issue of reversibility of effects, the suggestion that any diminished ability of DNA repair is temporary and disappears upon removal from exposure is questionable since there is limited evidence that an effect was present initially.

IX. MUTAGENICITY STUDY OF WORKERS EXPOSED TO ALKYLENE OXIDE (EO/PROPYLENE OXIDE) AND DERIVATIVES[187]

The purpose of this study was to investigate chromosome aberration rates among workers in a BASF factory which manufactures and processes alkylene oxides (including EO, propylene oxide, dioxane, epichlorohydrin, etc.). The study group consisted of 43 employees subdivided into four groups. Group I: employees with more than 20 years of exposure; Group II: employees with less than 20 years of exposure; Group III: "long-term" employees with accidental exposures; and Group IV: short-term, high-level exposure employees. The numbers in each group were 11, 6, 21, and 5 respectively. The control group consisted of 25 male medical department and office staff workers including 4 maintenance workers. Current exposure levels were estimated to have been less than 5 ppm with occasional peaks up to 1900 ppm for a duration of several minutes. Past exposures are estimated to have been "higher".

Only Group I was found to have had significantly increased chromosome aberration rates compared to controls. Excluding gaps, Group I had a rate of 3.5 compared with 1.36 for controls. A repeat examination of these same employees 10 months later revealed a reduced rate of 2.7%, although this remained significantly higher than that for the control group. Aberration rates for Groups II and III were somewhat elevated compared with controls (2.3 and 2.2%) but the differences were not statistically significant. Employees who had high but brief accidental exposures in the past had rates comparable to the control group (1.6%). The authors suggest the lack of adverse effects in exposure Groups II and III may be related to improved technology in more recent years which resulted in these employees reviving lower ("safer") doses. These results are interpreted as an indication that chromosomal aberrations may return to normal in accidentally exposed workers following their removal from EO exposure.

A. Critique

This study suggests that EO may have been the cause of some elevation in aberration frequencies seen in exposure Group I. The principal problems with this study are the small number of employees examined, the lack of precise exposure data and the fact that the employees studied were exposed to many chemicals other than EO including benzene, epichlorohydrin, and ethylene chlorohydrin. In order to determine representativeness of exposed and control subjects, it is essential to know whether the study was restricted to volunteers and, if so, what proportion of eligible subjects agreed to participate. Regarding exposures, the distinction between Groups I, II, and III is not sufficiently clear to enable any quantitative assessment of a dose-response relationship. In order to evaluate the full significance of the results for exposure Group IV, information on the length of time between the accidental exposure and chromosome testing is needed.

Other problems with this study include:

1. Uncertainty regarding whether the analyses of exposed and control workers were conducted "blindly".
2. The possible confounding effect of the slightly younger age of controls compared with exposed subjects.
3. The failure to control for extraneous factors such as smoking, drinking and drug-taking habits that are also related to elevated aberration frequencies.
4. The possibility that results may have differed if "median" rather than "mean" numbers of chromosomal aberrations had been compared.

X. MORTALITY AMONG EO WORKERS[75]

This study examined the mortality experience of 767 male employees exposed to EO at a Texaco chemical plant in Port Neches, Texas. Employees studied worked at least 5 years between 1955 and 1977, and mortality experience was traced through December, 1977.

Mortality experience was compared with the U.S. population adjusting for age, race, and time period. The plant had been in continuous production of EO since 1948. Based on an industrial hygiene survey conducted in 1977, the authors report that typical exposures to EO were less than 10 ppm. However, tank-car loading operations showed exposures at 6000 ppm.

A total of 46 deaths occurred in the exposed population compared with 80 expected (SMR=58). Of these, 11 were attributable to malignant diseases compared with 15.24 expected (SMR=72). There were no leukemia deaths among exposed workers and only 20 deaths from circulatory diseases (SMR=64). No significantly increased risk from any disease was reported. SMRs greater than 200 were reported for cancer of the pancreas, bladder, and brain as well as Hodgkin's disease; none of these are statistically significant due to small numbers. The authors conclude that EO does not appear to be a strong leukemogen; however, they do not rule out the possibility that the elevated SMRs for other specific cancer sites may be due to occupational factors, including EO exposure.

A. Critique

This essentially "negative" study has several limitations. First, the cohort size was small and the statistical power was low (i.e., only a tenfold excess of leukemia deaths could have been detected). Secondly, no latent period was considered and no information was provided regarding the percentage of the cohort that had survived a minimum latent period (10 years). Analyses were not performed by job title, department, age categories, length of employment, or length of survival; however, it is likely that the study size was insufficient to permit such analyses.

Although the evidence is not sufficient to rule out the possibility that EO is a human carcinogen, neither does it confirm suggestions which have arisen from earlier human and animal studies.

XI. PRELIMINARY REPORT OF PILOT RESEARCH CHROMOSOME STUDY OF WORKERS AT SITES WHERE EO GAS IS USED AS A STERILANT[186]

This analysis is based on a preliminary report of a pilot research chromosome study on Johnson and Johnson workers exposed to a mixture of EO and carbon dioxide gas used for the sterilization of medical products. Chromosome aberration rates and SCE rates were determined for exposed and control workers at each of three plants representing high, medium, and low relative exposure ranges for the pre-1980 period. Estimated 8-hr TWAs at these plants were I: 1 ppm; II: 1 to 10 ppm; and III: 50 to 200 ppm. At plants II and III (the medium and high relative exposure plants) employees were examined a second time, 6 months after the first examination.

No increase in chromosome abnormalities or SCEs were found for exposed employees at plant I, the low relative exposure plant, where estimated TWA_s of sterilizer operators were generally less than 1 ppm. At Plant II, the medium relative exposure plant, sterilant operators were found to have a significant increase in SCEs compared with the control group. This was based on a mean SCE rate of 14 per 100 cells for 4 exposed operators and a mean of 11/100 cells for 19 control employees. At Plant III, the higher relative exposure plant, a significant excess of SCEs was found in a group of two operators. Their mean SCE rate was 33/100 cells compared with 12/100 cells for the 21 controls. Moreover, a dose-response relationship was observed for SCEs among the three potential exposure groups. SCE rates for the background (n = 21), low (n = 24) and high (n = 2) potential exposure groups were 12, 14, and 33 SCEs per cell, respectively. After 6 months of nonexposure, the dose-response pattern persisted with SCE rates of 12, 15, and 35 respectively in background, low and high potential exposure groups. Chromosome aberration data (complex aberrations) did not provide consistent results. Although a dose-response relationship was suggested in the rates for Plant III, these were based on too few numbers to enable meaningful conclusions.

A. Critique

Concerns with this preliminary report include the small number of employees examined and the methods used for testing statistical significance. The number of operators tested at Plants II and III were relatively small (four and two, respectively) and it is unclear how the reported differences could be significant based on such small numbers. Aside from the two operators in Plant III, mean SCEs for all groups fell within a range of 8 to 15/cell and such differences may not be "clinically" significant.

Other issues of concern include: (1) the procedures for "retests" and (2) control for potential investigator bias. Regarding procedures for retests, employees in Plant I were not retested at all; at Plant II retests were performed following continuous 6-month exposures and at Plant III retests were performed after 6 months of nonexposure. The different procedures used raise questions about the comparability and interpretation of results. The decision not to retest at Plant I, because of a low number of SCEs on initial examination, is unfortunate as the retest data could have been used to assess intersubject variability. Secondly, the decision to question the exposure status of the control subjects at Plant III, based on the high number of SCEs obtained, may have introduced "investigator bias" by requiring, in effect, the control group to have a low rate. This may have introduced a bias in the direction of finding a positive association when one may not exist.

The extremely wide range of the estimated exposure at Plant III, i.e., 5 to 200 ppm, limits the usefulness of this study in identifying a no-effect level of exposure or quantifying a dose-response relationship. No information is given regarding the approximate exposures for the two operators at this plant who showed the highest SCE rates in the study. The extent of control for confounding factors (e.g., smoking status) is unknown since no data were presented with respect to these variables. Because of these uncertainties and the unknown validity of SCEs as predictors of adverse health outcomes, the significance of this report is unclear.

XII. SCE IN ASSOCIATION WITH OCCUPATIONAL EXPOSURE TO EO[75*]

This study incorporates and extends a Johnson and Johnson[186] pilot research chromosome report of workers exposed to EO in connection with the sterilization of health care products. The follow-up deals only with further analyses of peripheral lymphocytes for SCEs rates and does not include additional data on chromosomal aberrations.

In the pilot study, chromosomal aberration rates and SCEs were determined for exposed and matched control workers at each of the three worksites where estimated exposure levels prior to May 1980 were known to have differed. Estimated 8-hr TWAs at these plants were Worksite I, less than 1 ppm; Worksite II, 1 to 10 ppm; and Worksite III, 5 to 200 ppm. (In the present report, estimated exposure levels at Worksite I were said to be 0.5 ppm TWA$_s$ and at Worksite II, 5 to 10 ppm TWA$_s$.) In addition, employees in the study at each worksite were classified as having a high potential or low potential for exposure to EO (Table 5, Chapter 9).

The mean number of SCEs per cell for a given subject at a given sampling time was analyzed. With the exception of Worksite I, for which there was no 6 month follow-up, all exposed and control groups were tested initially and follow-up sampling was done 6, 12, and 24 months after the initial test. A community control group was added for Worksite III at 6 and 24 months (Table 6, Chapter 9).

Several different analyses of variance were examined in the application of statistical methods to the data. Age, sex, and smoking habits of the subjects were considered as possible confounders, and interlaboratory variation associated with blood sample processing and reading of slides was also taken into account.

A. Results

At Worksite I, classified as a low exposure plant (0.5 ppm TWA$_s$), mean SCEs were uniformly low and there were no significant differences between potentially exposed groups or between exposed groups and controls.

For Worksite II, pair-wise comparisons between exposure groups indicated that the high potential exposed group had significantly higher mean SCE levels than the low potential exposed group initially and at 12 and 24 months follow-up, but not at the 6 month sampling time. Similarly, mean SCE in the high potential exposure group was significantly elevated over control values initially and at 12 and 24 months. The low potential exposed group was not statistically significantly higher than the worksite controls at any follow-up period.

At Worksite III, pair-wise comparisons demonstrated that the high potential exposure group exhibited significantly elevated SCEs when compared with either the low potential exposure group or controls at all sampling times. Mean SCEs in the low potential exposure group differed significantly from those in the worksite control group at the initial and 6 month sampling, but not subsequently; significant differences were observed between this group and the community control when the community control group was included (6 and 24 months). Interestingly, the worksite control values were significantly elevated over community control values at both follow-up periods.

Smoking, sex, and age were shown consistently to influence SCEs; however, when analyses were adjusted for these three confounders, overall conclusions were not substantively altered.

B. Critique

The major concern with the report is the same as noted for the Johnson and Johnson preliminary study, i.e., the small number of employees examined, particularly in the high potential exposure group at Worksite II and III (N=4 and 2, respectively). The disproportionately small number of persons in the high exposure groups makes it difficult to put too much weight on tests of statistical significance. Nevertheless, in a qualitative sense, it appears that workers who experienced a relatively high level of exposure to EO generally tended to have increased SCEs, except at Worksite I.

An interesting intercomparison is the examination of mean SCE values of the high potential exposure group at each worksite and consideration of their relationship to the estimated TWA exposure values for the three plants. The highest mean SCE values were seen in the two workers at Worksite III where the 8 hr TWA for sterilizer operators was estimated to be 5 to 200 ppm. Sterilization at Worksite III ceased shortly after initiation of the chromosome study. At Worksite II, where the TWA estimates were 5 to 10 ppm, mean SCE values in the high potential exposure group generally were elevated over control values, but not to the extent seen at Worksite III. At Worksite I, there was no significant variance among any of the groups examined, although the authors state that a typical operator at Worksite I experienced over twice the number of sterilizer openings than those at Worksite III. Average exposure at Worksite I was estimated to be 0.5 ppm TWA_s. It appears that operators experiencing relatively high intermittent exposures to EO did not exhibit increased SCEs in a setting where the cumulative TWA was low. Again, this observation is based on a very small sample size and intermittent exposure levels for the operators at the three sites were not reported.

The authors consider the most striking observation in the study to be the persistence of elevated SCEs in the two workers at Worksite III for as long as 41 months after initial examination and 24 months after cessation of EO sterilization at the plant. No explanation is offered for this finding. It should also be noted that SCE values for one of these operators, although consistently high, showed considerable fluctuation from one sampling time to the next.

The follow-up information contained in this report does not increase its usefulness in identifying a no-effect level of exposure or in quantifying dose-reponse, two limitations which were noted in review of the preliminary report. The update also does not contain data showing the extent of control for confounding variables although it is stated that age, sex, and smoking were considered. The update is useful in that the retests tend to confirm differences among the groups observed in the initial and 6 month samplings. Sample size remains the most serious limitation of the study.

XIII. SCE AS A BIOLOGICAL MONITOR FOR WORKPLACE EXPOSURE: A COMPARATIVE STUDY OF EXPOSURE-RESPONSE TO EO[259]

The objective of the human study was to determine whether SCE analysis is useful as a biological monitor for the detection of EO exposure in hospital sterilization workers. For this purpose, sterilant workers at two hospitals in the San Francisco Bay area were monitored simultaneously for EO exposure (during typical working conditions) and SCEs. Measurements taken during the course of the study revealed that both hospitals were within the current 50 ppm TWA, OSHA standard. Although the 14 workers from the two hospitals were self-selected, a participation rate of 70% overall and 100% for those most highly exposed was attained at both hospitals. There were 13 controls selected from the clerical and administrative staff of the participating hospitals and two other research institutions. Comparison of habits between exposed and nonexposed workers revealed similarities with respect to smoking, caffeine intake, alcohol consumption, immunizations and infections, and drug taking. SCE measurements were based on a sample of 50 cells (cultured lymphocytes in peripheral blood) per worker. Exposed workers were divided into low and high exposure groups based on a determination of a cumulative exposure dose of less than or greater than 100 mg/m³ during a 6-month period.

The mean number of SCEs per cell for the high exposure group (n = 9) was 10.7 compared with a mean for the low exposure group (n = 5) of 7.8 and a rate of 7.6 for the control group. The increased frequency in the high exposure group was statistically significant (compared with controls); however, there was no statistical significance between mean values for the low exposure group and controls. Analysis by the Mann-Whitney Test revealed a significantly higher frequency of SCEs in all 14 exposed workers compared with controls. This study also demonstrated elevated SCE rates for smokers in both exposed and nonexposed groups. In exposed and nonexposed smokers, the group mean SCE rates were 9.09 and 7.89 compared with rates of 8.27 and 7.04 for exposed and nonexposed nonsmokers respectively. Based on these investigations, the author concluded that an increase in SCE rates can be detected in humans exposed to EO under normal conditions of work when exposures are determined accurately.

A. Critique

One of the concerns with this study was that means rather than medians were used to compare exposed and nonexposed SCE frequencies. When median SCE values are compared, there is little difference between the median SCE rate for exposed workers (8.2) vs. the median for control workers (7.5). In general, the differences reported in this study are extremely small and the number of subjects investigated too few to permit extrapolation of the results of this study to general populations.

XIV. SPONTANEOUS ABORTIONS IN HOSPITAL STAFF ENGAGED IN STERILIZING INSTRUMENTS WITH CHEMICAL AGENTS[86]

This is a report of a study of the incidence of spontaneous abortion among female sterilization workers in Finland which was initiated in mid-1979 by the Finnish Institute of Occupational Health. Its purpose was to evaluate the effect on reproduction of occupational exposures to three sterilant agents used in Finnish hospitals, EO, glutaraldehyde, and formaldehyde. A group of 1026 sterilization workers and a control group of 1051 nursing auxiliaries were identified by members of the supervisory nursing staff at each of 80 hospitals in 1979. Each identified worker in the study and control groups was sent a questionnaire which asked for information on past reproductive and employment experience, as well as other demographic and lifestyle information. The response rate for each group was above 90%. Respondents who had never been pregnant or were never married were excluded from the study. The final study groups consisted of 645 responding sterilizer operators and 574 responding control workers. Each reported pregnancy for sterilizer operators was classified as exposed or nonexposed to the above three sterilizing agents based on the occupation reportedly held at the start of pregnancy. (Jobs had been previously associated with specific sterilization agents by supervisory nurses at each hospital.) For pregnancies classified as exposed to each agent or combination of agents, the spontaneous abortion rate was computed using all pregnancies and their outcomes reported by the women themselves (through 1981) as well as all hospitalized pregnancies and their recorded outcomes during the 6 year period 1973 through 1979.

Spontaneous abortion rates for exposed pregnancies were calculated and compared to rates for nonexposed pregnancies using logistic regression analysis to adjust for maternal age, parity, decade of pregnancy, smoking habits, and drinking (alcohol and coffee) habits.

Typical exposures to EO were reported as 0.1 to 0.5 ppm TWA$_8$ with peaks to 250 ppm, based on sampling data which had been collected by the Institute since 1976.

A. Results

The results of this study indicated that exposed sterilizing staff pregnancies resulted in higher spontaneous abortion rates than either the nonexposed pregnancies to the same women or the pregnancies of control workers. For all three groups of pregnancies, rates of spontaneous abortion (from questionnaire data) increased over time, indicating that the outcomes of recent pregnancies were likely better recalled than the outcome of early pregnancies. Pregnancies exposed to EO alone or in combination with other sterilant agents had spontaneous abortion rates which were significantly higher than those for pregnancies not exposed to EO (16.1 vs. 7.8%). Analyses based on hospitalized pregnancies between 1973 and 1979 revealed a higher rate for EO exposed pregnancies (22.6%) than either nonexposed pregnancies to the same women (9.9%) or pregnancies among control staff (9.2%).

B. Critique

Numerous methodologic and analytic problems limit the confidence which can be placed in the results of this study. The principal methodologic problems include the possibility of "recall bias" introduced by the survey instrument, the inadequate measurement of exposure to EO or the other two sterilant agents studied, and the inadequate validation of self-reported spontaneous abortions.

1. Recall Bias

Most retrospective interview studies are subject to problems in accurate recall of either the exposure or the effect.[86a] In this study, ascertainment of both exposure status and pregnancy outcome was subject to errors in recall on the part of study subjects. "Blinding" of study subjects to keep them uninformed as to whether they are part of the exposed or unexposed group is generally considered essential to preclude differential recall and reporting of the effect of interest. In this study, the questionnaire which was used differed in color for sterilant staff and control staff workers and also contained information which informed study subjects that the purpose of the study was to investigate adverse reproductive effects associated with sterilization gases. (Of the three agents studied, only EO is used as a gas, so that study subjects may have been alerted specifically to this exposure.)

The results of this study clearly indicate that recall bias was associated with the timing of the pregnancy, so that more recent pregnancies were recalled more accurately than earlier pregnancies. Because exposed pregnancies to the sterilization staff were concentrated in the more recent years, there was likely to be a differential recall bias effect for exposed and nonexposed pregnancies which could have resulted in a higher abortion rate for exposed pregnancies.

2. Exposure Data

At the start of the study in 1979, nurses at each hospital were asked to compile a listing of all occupations which involved exposure to one or more sterilization agents at their hospital. No historical record of exposure or exposure levels was available to assist the nurses in this task. The fact that this was done retrospectively introduces the possibility of some recall error on the part of the hospital nurses. The additional fact that the three agents under study had been introduced at different points in time, with EO having been introduced most recently, raises the possibility that classification errors were more likely to have occurred for glutaraldehyde and formaldehyde than for EO.

Pregnancies (from 1950 to 1981) were then classified as exposed or nonexposed to one or more sterilization agents based upon the occupation (reported in the questionnaire) held at the start of each pregnancy. A dichotomous classification of "exposed" or "not exposed" was used to designate whether a pregnant employee had worked in a department that used one of the three sterilant agents in question. No method was used to validate the accuracy of the respondents' recall of their past occupations or the timing of their pregnancies with respect to those occupations. Because EO was the newest of the three sterilant agents used in Finnish hospitals, and because sterilization gases were specifically mentioned in the cover letter to study subjects, it is likely that EO exposed pregnancies were recalled more accurately than other exposed pregnancies.

3. Validation of Effect

The self-reported incidence of spontaneous abortion is subject to problems of reliability and validity. The historical occurrence of spontaneous abortion, or miscarriage, may be easily forgotten, if it occurred in the distant past and was followed quickly by a successful pregnancy. Early occurrences of spontaneous abortion (first trimester) may go undetected or mistaken for a missed menses if medical attention is not sought. Alternatively, delayed or missed menses can be easily misdiagnosed by the individual herself as a spontaneous abortion or miscarriage. Although some information on hospitalized cases of spontaneous abortion was available, it did not cover the spectrum of pregnancies reported in the questionnaire, nor were these data used for the purpose of validating self-reported cases.

Of some concern in this study is the possibility that knowledge of the study's hypothesis regarding EO led to an increased reporting of spontaneous abortions among exposed pregnancies. Although the results of the analyses which were based on hospitalized pregnancies would be free of reporting and recall bias, the number of EO exposed pregnancies between 1973 and 1979 was too small to provide confirmation of the questionnaire data.

4. Analytic Problems

The presence of serious analytic problems is suggested by inconsistencies in the reported results based on the questionnaire data and the comparison of those findings with the results based on the hospital data.

Examination of Table 1 illustrates the inconsistency of results based on the questionnaire data for nonexposed sterilizing staff and control staff pregnancies.

Table 1
SPONTANEOUS ABORTION RATE AMONG STERILIZING STAFF AND NURSING AUXILIARIES (CONTROLS)

	Total No. of pregnancies	Spontaneous abortion rate	
		Crude (%)	Adjusted (%)
Sterilizing staff	1443	11.3	9.7
Exposed during pregnancy	545	16.7*	15.1*
Exposure uncertain	293	12.3*	11.3*
Not exposed during pregnancy	605	6.0	4.6
Controls	1179	10.6	10.5

Note: Rates adjusted for age, parity, decade of reported pregnancy, coffee and alcohol consumption, and smoking habits.

* $p < 0.001$ (exposed vs. nonexposed pregnancies).

From Hemminki, K. et al., *Br. Med. J.*, 285, 1461, 1982. With permission.

It can be seen from this Table that adjusted rates for sterilizing staff and control pregnancies are very similar. However, the rate of spontaneous abortions among nonexposed sterilization workers (4.6%), which should be equivalent to that among control workers (10.5%), is substantially lower.

One likely explanation for the extremely low rate of spontaneous abortions among nonexposed sterilant staff pregnancies is that they occurred in the earlier years covered by the questionnaire survey and were consequently at lower risk of abortion due to younger maternal age and lower parity. Also, these pregnancies were likely to be most affected by recall bias. Although the authors claim to have adjusted their rates for age, calendar time period, parity, and other factors, it is possible that the methods used were not sufficient to enable comparability of rates among these disparate groups. (The usefulness of the data on smoking and drinking habits, which related to the woman's "current" habits and not to habits at the time of pregnancy, must also be questioned.)

Figure 1 further illustrates this inconsistency by demonstrating that the differences among the three groups persisted throughout the time interval covered by the study.

Although this increasing rate over time is interpreted by the authors as indicative of better recall of pregnancy outcomes for the most recent decade, no interpretation is offered regarding why the nonexposed sterilizing staff had lower spontaneous abortion rates than controls for all three decades.

Table 2 from the report contains results specific to each of the three sterilizing agents ("crude" rates have been omitted).

In this analysis, the nonexposed rates represent a combination of pregnancies of the nonexposed sterilant operators and all pregnancies of the control workers. In addition, for any specific agent, the nonexposed rate contains pregnancies that were exposed to any other agent. The appropriateness of including pregnancies exposed to any sterilant agent in the nonexposed rate is highly questionable in a study of this type. In view of the abnormally low rate obtained in Table 1 for nonexposed sterilant staff workers, the nonexposed rates contained in Table 2 are likely to be subject to the same problems of interpretation and an exposure-specific bias cannot be ruled out. In general, the unorthodox and unusual manner in which these data have been analyzed detracts from the overall credibility of the results obtained.

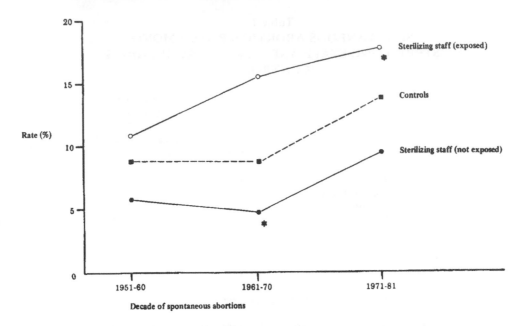

FIGURE 1.   Parity-adjusted frequency of reported spontaneous abortions during three decades in three groups of women aged 20 to 34 years. (*$p < 0.05$ from the mean rate (10.6%) for the controls.)

Table 2
EFFECT OF EXPOSURE TO EO, GLUTARALDEHYDE, AND FORMALDEHYDE ON THE FREQUENCY OF SPONTANEOUS ABORTIONS

	Not exposed		Exposed	
Chemical sterilizing agent	No. of pregnancies	Adjusted rate (%)	No. of pregnancies	Adjusted rate (%)
EO (with and without other agents)	1004	7.7	146	12.7[a]
Glutaraldehyde (with and without other agents)	710	7.7	440	9.3
Formaldehyde (with and without other agents)	1100	8.3	50	8.4
EO (with glutaraldehyde)	704	6.6	446	11.3[a]
EO alone	1068	7.8	82	16.1[b]
Glutaraldehyde alone	786	7.8	364	9.4

Note: Rates adjusted for age, parity, decade of pregnancy, smoking habits, and alcohol and coffee consumption.

[a] $p < 0.005$.
[b] $p < 0.01$.

From Hemminki, K. et al., *Br. Med. J.*, 285, 1461, 1982. With permission.

The hospital discharge data base consisted of approximately 374 pregnancies for the sterilizing staff and 368 for the control group for the years 1973 to 1979. Table 3 is reproduced from Hemminki et al.[86]

It can be seen that the rate and ratio of spontaneous abortions for EO are based on only 7 spontaneous abortions and 31 total pregnancies. Although the authors interpret

Table 3
DETAILS OF SPONTANEOUS ABORTIONS AMONG STERILIZING STAFF AND NURSING AUXILIARIES (CONTROLS) FROM 1973 TO 1979

	Total No. of pregnancies	Spontaneous abortions	
		Rate[c]	Ratio[d]
Sterilizing staff			
Exposed to:	253	14.2[a]	19.3[b]
EO	31	22.6[c]	33.3[a]
Glutaraldehyde	178	12.9	17.3
Not exposed	121	9.9	12.6
Controls	368	9.2	11.8

Note: Data obtained from hospital discharge register.

[a] $p < 0.05$ when compared with controls (Chi-square tests).
[b] Not significant ($p < 0.1$).
[c] Number of spontaneous abortions/number of pregnancies.
[d] Number of spontaneous abortions/number of births.

From Hemminki, K. et al., *Br. Med. J.*, 285, 1461, 1982. With permission.

these results as supporting the findings based on the questionnaire data, the numbers are clearly too small to provide any meaningful validation or confirmation of an association between EO exposure and increased risk of spontaneous abortion.

It is of interest to note that Table 3, which is based on hospital discharge data for the most recent 8-year period, does not substantiate the difference between the spontaneous abortion rates of nonexposed sterilizing staff and control workers. The comparability of rates between these two nonexposed groups lends confirmation to the suspicion that the methods used in analyzing the questionnaire data may have been seriously flawed.

Because of the number of problems with this study, both in design and analysis, the results are difficult to interpret, at best. Further studies on this subject are clearly needed.

XV. MORTALITY STUDY ON EMPLOYEES EXPOSED TO ALKYLENE OXIDES (EO/PROPYLENE OXIDE) AND THEIR DERIVATIVES[182]

The purpose of this study was to evaluate the mortality experience of a cohort of 602 employees exposed to alkylene oxide and a number of other substances at eight production plants where alkylene oxides and their derivatives are produced. The study cohort consisted of active and former employees who had worked a minimum of 6 months between 1928 and June 30, 1980. The observation period for this mortality study covered the 52-year period from 1928 until June 30, 1980. Indirectly standardized mortality ratios (SMRs) were calculated for the cohort using four different comparison populations:

1.   Ludwigshafen (180,000)
2.   Rhinehessia — Palatinate (1.8 million)
3.   The Federal Republic of Germany (60.5 million)
4.   An internal cohort comparison group of styrene plant employees (1,662)

Overall results were presented and interpreted. In addition, causal relationships were investigated by examining latency (10 years) in a specific subgroup as well as lengths of employment in a specific subgroup.

The mortality experience of this employee cohort failed to indicate an increased risk of cancer or other causes of death associated with exposure to alkylene oxides. In the overall cohort, 56 deaths were observed vs. 76.5 expected and 12 malignant neoplasms were observed compared with 15.5 expected. The results for specific cancers were 4 deaths due to stomach cancer (2.7 expected) and 2 deaths from hematopoietic and lymphatic organs (1.1 expected). These results, which were not significantly different, were based on a total of 8484 person-years of observation. A subgroup of 351 employees (3,779 person-years) with a minimum observation period of 10 years was also investigated. The only significant increase found in this cohort was deaths due to diseases of the urinary organs and genitalia — 3 observed vs. 0.98 expected. There were 10 cancer deaths observed vs. 11.8 expected. There were 2 stomach cancer deaths observed vs. 2 expected; and 1 death attributable to cancer of the hematopoietic lymphatic organs vs. 0.75 expected. The authors concluded that the analysis of changes in mortality ratios (O/E) with increasing years of employment failed to suggest an occupational etiology for any specific cause of death.

A. Critique

One limitation of this essentially negative study is the relatively small sample size (n = 602) and its lack of statistical power for detecting an increased risk if one does exist. The two technical problems with the study include potential of bias in determining cause of death and insufficient control for temporal trends in determining expected mortality rates. With regard to the first problem, the author states that consultation with personal physicians was used to supplement information obtained from the death certificates. This may create a problem if this type of adjustment was not done for the comparison area death certificates which were used. It is difficult to know how this might have affected the results since no information is given regarding how the diagnoses were "changed" as a result of this information. The second problem regards the fact that rates for the single time period 1970 to 1974 were used to generate expected numbers of deaths in the study cohort between 1928 and 1980. This failure to control for time could be a problem if there were significant increasing or decreasing trends over this time period for any of the diseases under investigation. In general, however, this study appears to have been well conducted and the authors' interpretation appears to be reasonable.

Appendix B

ISSUES RELATING TO INHALATION CARCINOGENICITY STUDIES ON ETHYLENE OXIDE

I. INTRODUCTION

The differences between the results of the NIOSH study and those of the BRRC study are noteworthy. Taking mononuclear cell leukemia (MCL) as an example, the incidence in male control rats was as follows: NIOSH, 33%; BRRC (two groups), 10 and 16%. After 2 years of EO exposure at 100 ppm (NIOSH rats were exposed for 1 hr longer each day), NIOSH reported 67% MCL, while the corresponding group at BRRC had 30% MCL. Consideration needs to be given to possible explanations for this difference, as well as other issues pertinent to these and other similar studies.

The conduct of the BRRC study met or exceeded regulatory requirements and has passed the test of peer review by scientists within industry, government, and academia. The following discussion deals with facets of the study which complicate its use in supporting an assessment of cancer risk to human populations potentially exposed to EO.

II. EPIZOOTIC INFECTIONS

The sialodacryoadenitis (SDA) virus infection appearing in the BRRC study in all exposure groups (including controls) began in the 15th month of the study and was characterized by loss in body weight, conjunctivitis, proptosis, and enlarged salivary glands. Mortality among females was greater than in males, and the 33 and 100 ppm exposure groups had a higher mortality than the 10 ppm and control groups. Histopathologic examination of the animals which died during the peak of the infection revealed no specific cause of death. Exposure to EO was discontinued. After 2 weeks, clinical signs disappeared, body weights increased, and mortality declined, whereupon exposures were resumed after the 2-week respite. Mortality declined after the 17th month. In order to deal with the increased mortality that had occurred in the two higher dose groups, it was decided to recalculate the data on the basis of the number of animals alive at the beginning of the 17th month. When the mortality began to increase again in the 20th month, there was no apparent reason to associate the late treatment-related, differential mortality with the previous viral infection. The differential mortality due to combined viral and treatment effects prior to month 17, and the differential mortality due to EO exposure after month 17, resulted in only 29 males and 26 females surviving to the end of study. (In the NIOSH study only 15 males of the 100 ppm group survived to 24 months.)

Epizootic infections are a complication of long-term animal studies which cannot be absolutely avoided, even with the best of husbandry. The presence of an epizootic infection might be evidenced by conjunctivitis, labored breathing or rales, and by a transient depression in weight gain over some brief period. As the technology has become more accessible and practical, it is now common in modern chronic studies to attempt identification of the epizootic agent. For example, SDA virus has been identified as the agent responsible in recent studies on formaldehyde and on methyl chloride. The NIOSH chronic inhalation study on EO suffered a *Mycoplasma* infection in the rats, and the Dunkelberg gavage study[144] was complicated by an epizootic infection which gave rise to pneumonia. In each of the instances cited, the extent to which such epizootic infection jeopardizes the entire study, or merely influences the final outcome,

merits careful consideration. It is important also to understand the nature of the interaction, if any, between the infectious agent and the test material. This would enhance
the reliability of the otherwise uncomplicated chronic studies for purposes of application in human risk assessment. In the particular case of EO, the SDA virus appeared
after 15 months of EO exposure, rather than in young animals (3 to 5 months). In
addition, sex-related (BRRC study) and treatment-associated differences in mortality
(both BRRC and NIOSH studies) suggest an effect on the immune status of the Fischer
344 rats. Normal immune defenses may have been depressed, or (in the case of the
virus) a latent virus infection may have been activated by EO exposure. Although SDA
virus has not been associated with the development of chronic disease in rats, other
types of viral infection are known to give rise to leukemia in mice and chickens.

III. SIGNIFICANCE OF TUMOR TYPES TO MAN

In the BRRC and NIOSH studies, increased incidences or rates of development of
MCL were observed. Tumors of multiple organs and types were reported in the BRRC
study, with increased incidence of peritoneal mesotheliomas (PM) in the males and
earlier development of pituitary adenomas in males and females. An increased incidence of primary brain neoplasms was reported in the BRRC study and of gliomas in
the NIOSH study.

MCL is an uncontrolled proliferation of a particular type of white blood cells. These
cells (abnormal lymphocytes) appear in the peripheral blood in very large numbers.
They frequently appear in the spleen and often in the liver. The disease occurs spontaneously in Fischer 344 rats and no corresponding lesion is found in other strains or
stocks of rats and mice. Thus, MCL is a major cause of death in senile F344 rats. The
leukemia appears to originate in the spleen and spreads to other organs.[238]

A number of reasons may account for the difference in incidence of MCL between
the NIOSH and BRRC studies. The sources of the Fischer 344 rats were different
(NIOSH rats were from Harlan Industries, Indianapolis, Indiana, while the BRRC rats
were from Microbiological Associates, Walkersville, Maryland), and the diets (specified in the studies) were different. Both factors may contribute quantitatively and qual-
itatively to differences in the spontaneous tumor incidence. Time of death is also an
important factor: as mentioned above, MCL is a late-occurring tumor, its incidence
increasing exponentially as the age of the F344 rat increases. The tumor incidence in a
group of these animals is, therefore, very much dependent on the time at which those
animals are killed (24 months in the NIOSH study and up to 26 months in the BRRC
study).

Human leukemias are a heterogeneous group of white blood cell diseases which are
broadly grouped into the myelocytic and lymphocytic types. According to R. R. Maronpot,[239] some forms of human acute myelogenous leukemia have several features in
common with MCL of Fischer rats. Thus, Dr. Maronpot considers that, while MCL
of Fischer rats is "unique," it is not entirely different from other types of leukemia.
Although many of the lymphocytes which proliferate in MCL morphologically resemble normal rat large granular lymphocytes (LGL),[238,238a-e] the leukemic cells are char-
acteristically heterogeneous, with differences in cytotoxic activity and cell surface antigens. The normal human lymphocyte population contains a subpopulation of LGLs
which originate in the bone marrow. This specific cell type rarely proliferates in human
leukemic diseases. The literature contains a limited number of case reports of patients
with acute or chronic lymphocytic leukemia that consisted of a homogeneous population of cells with properties similar to those of LGL.[240,241] Despite the presence of some
common features between the so-called human chronic T_γ-lymphoproliferative
disease[238e] and LGL leukemia in Fischer 344 rats, there are substantial clinical differ-

ences between the two conditions. Reynolds and Foon[238e] consider the two to be morphologically, functionally and clinically similar. They refer to unpublished observations suggesting that the selective development of LGL leukemias in Fischer 344 rats "may reflect the susceptibility of LGLs and LGL precursors to transformation by an oncogenic retrovirus." Stromberg[238f] suggests that large granular lymphocyte leukemia in F344 rats can serve as an animal model for human T_γ-lymphoma, malignant histiocytosis and T-cell chronic lymphocytic leukemia. He admits, however, that "the relationship of these neoplasms ultimately depends upon identification of cell type", and proposes that development of surface antigen profiles for the rat MCL should permit comparison with differentiation antigens on human leukocytes. When more research is completed, better understanding may be achieved of the differences in tissue of origin, population homo- or heterogeneity, and spontaneous incidence, as well as the effects of chemical carcinogens on the incidences of these leukemias. The conclusion to be drawn is that it is wrong at present to equate MCL in rats with the overwhelming majority of human leukemias.

Not surprisingly, therefore, many pathologists believe that there is no exact correlation between any common form of human leukemia and MCL of F344 rats. This weakens the extrapolation of the MCL to any specific human disease like the "leukemias" reported by Högstedt et al.:[181] chronic myeloid leukemia, acute myelogenous leukemia, and Waldenstrom's macroglobulinemia (see Chapter 10 for an analysis of this work). The increased incidence of MCL in both the NIOSH and BRRC studies may reflect the carcinogenic potential of EO, and thus provide a quantitative basis for human assessment of hazard. However, until more is known about both the pharmacokinetics and mechanism of action of EO, it is premature to conclude that EO is a human leukemogen.

PM is a tumor which arises in the single celled mesothelial lining of the abdominal cavity. Its incidence is experimentally increased by nitrosopyrrolidine or asbestos.[242] The tumors in the BRRC study arose on the tunica vaginalis of the testis, the common site of origin of this tumor. Although the mesothelium covers the male reproductive organs inside the abdomen, it is not a part of the reproductive system. The background or spontaneous incidence (about 2%) of PM in the F344 rat is much lower than MCL or pituitary adenomas. In rats exposed to EO, the increased incidence of PM occurs late in life, unlike the early induction of these tumors by potent chemical carcinogens. This "death bed" appearance of PM suggests that EO may be modulating the natural late expression of this particular tumor.

Gliomas are microscopic tumors most frequently found in the cerebrum of aged rats. The reports of the NIOSH and BRRC studies showed an increased incidence of gliomas and primary brain tumors, respectively, in comparison with the low incidence of such tumors in controls. All of these tumors were found to occur in the later portion of the animals' lifespan.

The criteria to be evaluated, in reaching a decision that a chemical agent is acting as a neurocarcinogen, are as follows:[243]

A. Establishment of a Dose-Response Relationship

At first sight, this relationship seems to have been demonstrated. However, the apparent incidence of brain tumors is related to a number of experimental factors: the ability of the tumors to be seen on gross examination at necropsy, the locations of tumors in the brain, the number of blocks which are taken from the brain at autopsy, the number of sections cut from each block and their locations, and the age of the animals.[244] In the NIOSH study, and in the BRRC study, three sections were taken: one in the cerebrum, one in the midbrain region, and one in the cerebellum. A number of additional sections were cut in the NIOSH study, and these were approximately randomly distributed between treatment groups.[245]

"It was disappointing to note that there was a marked lack of uniformity at the tissue (brain) trimming stage. Some rats had a single sagittal section of the brain, others a single transverse section of the cerebellum. Approximately half the rats from each group had three transverse sections (cerebellum-1, cerebrum-2), while a few rats had from two to eight sections of various parts of the brain. In a brief analysis of the number of section, there did not appear to be a particular sampling bias for a group, and it was the PWG's (Pathology Working Group's) opinion that the glioma incidence could not simply be explained by the sampling technique, however the lack of uniformity detracts from the study."

This factor should be verified systematically to insure no influence on the conclusions drawn regarding brain tumors. Most brain tumors are seen in the cerebrum, with a lower frequency found in the midbrain; hence, in effect only one section or at most two sections were taken in the most appropriate region of the brains. Based on the statement of Ward and Rice[244] (paraphrased above) and the criteria of Koestner,[246] it may be inferred that the reported incidences of brain tumors in all groups in the NIOSH and BRRC studies may be too low. In addition, there may be an effect on the apparent dose-response relationship due to the sampling procedure.

B. Decreased Latency of Tumors in Exposed Animals

Demonstration of an earlier appearance of neurogenic tumors and a decreased time-to-death is a hallmark of neurocarcinogenic action.[247-249] Strikingly, there was no discernible decrease in survival time with increased exposure to EO. The appearance of gliomas late in the lifespan of the exposed rats indicates that EO does not affect the latency of these neoplasms, just as it does not decrease the latency of other spontaneously occurring rat tumors.

C. Shift to More Anaplastic Tumor Types in Exposed Animals

No increase in the degree of anaplasia is discernible as a consequence of the action of EO.

D. Presence of Early or Preoplastic Lesions

The NIOSH but not the BRRC study has reported cerebral gliosis (Table 5, Chapter 8), unaccompanied by inflammation, neuronal degeneration, or necrosis. The suggestion was made that the gliosis represented incipient gliomas. However, the presence of glial nodules was not reported in the brains of any of the rats, as seen for example with neurocarcinogens such as the nitrosoureas.[250-252] The absence of such preoplastic lesions in EO-exposed rats is noteworthy.

E. Distribution of Induced Neurogenic Tumors Throughout the Central and Peripheral Nervous Systems

Contrary to experience with most neurocarcinogens[253] neurogenic tumors in rats exposed to EO occurred only in the central nervous system.

F. Higher Susceptibility of Fetal Brain Cells than Adult Brain Cells to Neurocarcinogenic Action

Transplacental exposure to EO has not been studied in relation to neurocarcinogenicity. Ward[158] has speculated, in the absence of experimental support, that EO may cause a much higher frequency of brain tumors if given transplacentally, based on a comparison with a very potent carcinogen, ethylnitrosourea (ENU). Ward's supposition may be related to the superficial similarity of anticipated DNA adducts formed by EO and ENU: hydroxyethyl — compared to ethyl — substituents are the demonstrated reaction products with cellular macromolecules. However, the presence of the hydroxyl group may greatly alter the highly specific activity of enzymes involved in DNA transcription, replication, and repair.

Pituitary adenomas are very common in the Fischer 344 rat, and although there exists a human counterpart, there was no increase in incidence of the adenoma in the BRRC study, and in the NIOSH study a dramatic reduction occurred in the EO-treated groups. Hence one can assign a low order of importance for this tumor in terms of utility for human risk assessment.

Finally, the choice of an animal model predetermines an inability to detect certain types of toxic endpoints. In the male F344 rat, the incidence of testicular interstitial cell tumors is virtually 100% at 2 years of age. This makes it infeasible to detect late effects of test materials on the testes in chronic studies. In trying to select the most relevant tumor types detected in the chronic EO inhalation studies for human risk assessment, some may be omitted due to inherent limitations of the model itself.

IV. QUANTITATIVE ANALYSIS OF THE DATA (SEE ALSO APPENDIX C)

The data in the BRRC study were evaluated by life table analysis and by mortality-adjusted trend analysis at the end of the study. Although it may be argued that bias is introduced by deviation from the statistical methods in the original protocol, which did not include either of these tests or the use of the Bonferroni correction factor, the tests were introduced for consistency with the techniques applied in the NCI bioassay program at the end of 1980. Enough raw data was included in the final report to permit other types of statistical analysis to be applied.

One of the major problems in evaluating data from any chronic study is in making adjustments for loss of animals during the course of the study. One may evaluate data from the terminal kill and observe the effect of exposure on incidence of tumors; however, the number of animals surviving to this point is different for each exposure group. In fact, mortality was affected by exposure in a dose-related manner in the BRRC study. The techniques employed in that study are an attempt to control for this difficulty by examination of the time course of tumor development (i.e., discovery) using the calculated number of animals at risk during set time intervals. This approach introduces another uncertainty. Some tumors are discovered in animals dying or killed moribund and the pathologist must decide whether or not the tumors are "life-threatening" or "nonincidental." Other tumors are incidentally discovered in animals at interim kills. If these latter animals are not permitted to live long enough to die from the tumor, its premature discovery should not be accorded the same importance as that same tumor type which kills another animal. Methods selected in the BRRC study were one attempt to deal with this dilemma: the number of animals at risk in each group was calculated at monthly intervals for all animals alive in that group at the start of the month. The number at risk was adjusted for animals found dead, killed moribund, or killed at scheduled interim times or at study termination.

Life table analysis employs a final interval, in the time scale of the BRRC study, of 24 months to infinity. Since all animals were sacrificed by the 26th month, premature discovery of some tumors occurred. These tumors may not have been detected (at least in theory) if the animals had lived out their entire lifetimes.

The sensitivity of the tumor incidence to the time-of-death of the animals is crucial when the tumor is common in aging rats and is exponentially approaching 60 to 100% around 2 years of age. For example, in the BRRC study, a difference in the incidence of pituitary adenomas would have been observed if the study had terminated 2 months earlier. Similarly, it may be argued that the MCL incidence may have been equivalent between treated and control animals if the study had continued 2 months longer. The latter suggestion is hypothetical and unlikely, but illustrates the point that observed differences in tumor incidence of very common spontaneous tumors in senile rats do not carry the same weight in an assessment of compound-related effects as the appearance of unique tumors occurring earlier in life. The conclusions of the BRRC study are, therefore, complicated by the absence of a unique early-occurring tumor. The NIOSH study supports this position. Nonetheless, these chronic EO inhalation studies are the only studies employing a route of administration which is relevant to the predominant form of human exposure. Since the BRRC study was conducted on a larger scale, provides much more information, and the results are in general concordance with those of the NIOSH study, it is appropriate to choose the BRRC study for use in risk assessment.

Appendix C

IMPLICATIONS OF THE TIME TO RESPONSE INFORMATION

R. L. Sielken, Jr.

I. TIME TO RESPONSE ANALYSES NOT INVOLVING MODELING

A scientific hazard assessment should evaluate all of the data. There is a great amount of information concerning the time at which a response to EO vapor might be observed. The final report on the Bushy Run Research Center (BRRC) study of EO inhalation in rats contains several figures and tables as well as two detailed appendices containing time to response information.[27] In this Appendix several implications of this information will be explored.

Figures 1 to 3 indicate the cumulative proportion of rats developing a specified response (death with peritoneal mesothelioma in male rats, death with mononuclear cell leukemia in male rats and death with mononuclear cell leukemia in female rats, respectively) plotted against the number of months of exposure. (These figures are based on the life table analyses in the BRRC final report[27] — see BRRC Table 16 (Chapter 11) and BRRC Figures 6 to 8 therein.) There are at least two conclusions which can be drawn from these figures. The first is that, if a response occurs at all, it occurs very late relative to an average rat lifetime. The second conclusion is that the number and timing of the responses at 0 and 10 ppm are very similar. This latter conclusion is made even more evident in Figures 4 to 6 which are the same as Figures 1 to 3 except that the cumulative proportion of rats developing a specified response is plotted only for the 10 ppm exposure group and the two control groups.

There are several other ways to characterize the lateness of the animal response times, the similarity between the response times for the rats exposed to 10 ppm and response times for the control rats, and the changes in the response times at 33 and 100 ppm relative to those at 0 and 10 ppm.

For example, Table 1 indicates the number of months in the experiment until the proportion of animals with the designated response reaches specified levels. In particular, when the response of interest is mononuclear cell leukemia in female rats, it took 23 months before at least 5% of the female rats in the control group developed the response, 24 months for the 10 ppm group, 20 months for the 33 ppm group, and 21 months for the 100 ppm group. Throughout Table 1 the rats at 0 and 10 ppm took similar times to reach the same response rate (risk level); in fact, it never took less time to reach a specified risk level at 10 ppm than it did at 0 ppm.

The cost of being exposed to EO can also be considered in terms of the length of lifetime without a toxic response. The estimated mean number of experimental months without a toxic response can be calculated using the information in Table 16 (Chapter 11) of the BRRC final report.[27] In the BRRC study the estimated mean would be 25 months if no rat at a particular dose level developed the response and would be 20 months if half of the rats developed the response at the end of the 15th month and the other half did not develop the response during the 25 month experiment. For a particular definition of response (e.g., mononuclear cell leukemia or peritoneal mesothelioma) the calculation would be

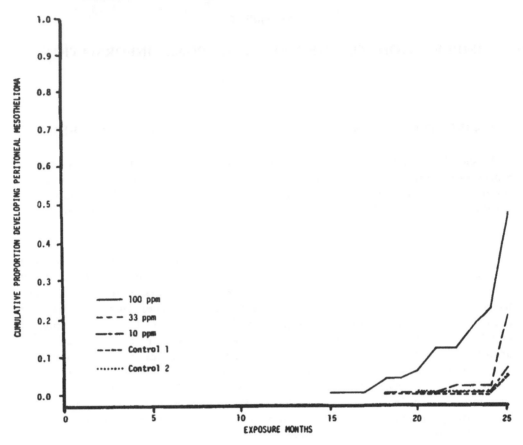

FIGURE 1.   The cumulative proportion of male rats at all exposure levels that developed peritoneal me-sothelioma in the BRRC EO inhalation study.

$$\text{Mean number of months without a response} = \sum_{m=1}^{25} (m - 1)\text{Prob(response develops at beginning of month m at the given dose)}$$

$$+ 25 \text{ Prob(response doesn't develop during the 25 month experiment at the given dose)}$$

where the probabilities (Prob) are estimated from the cumulative percentages in BRRC Table 16 (Chapter 11). (A rat tabled as developing a response during a month is assumed to have developed that response at the beginning of that month.) The calculated mean number of months without a response developing is given in Table 2 for the four experimental dose levels (0, 10, 33, and 100 ppm) and the different definitions of a response. Table 2 also indicates the corresponding mean percentage of the experimental period without a response. The information in Table 2 clearly suggests that the mean time without a response is virtually the same at 0 and 10 ppm and hardly decreased even at 33 ppm. Since the potential time of a response is so important to a subject exposed to a hazard, it definitely seems appropriate to consider time to response summaries such as those in Table 2.

The survival curves corresponding to the times without specified responses can be tested for equality using the statistical procedures described in Tarone.[263] (The survival curves correspond to 1.0 minus the cumulative incidence curves shown in Figures 1 to 3.) When the time to response data in Table 3 was tested, the null hypothesis that all

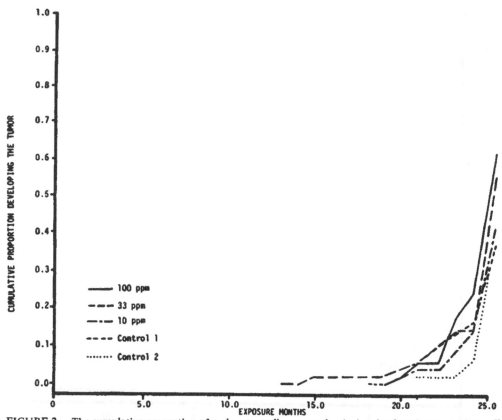

FIGURE 2. The cumulative proportion of male rats at all exposure levels that developed mononuclear cell leukemia in the BRRC EO inhalation study.

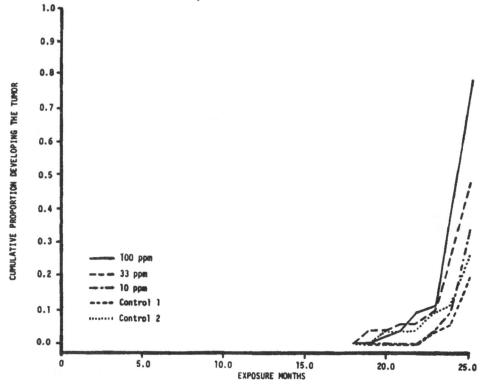

FIGURE 3. The cumulative proportion of female rats at all exposure levels that developed mononuclear cell leukemia in the BRRC EO inhalation study.

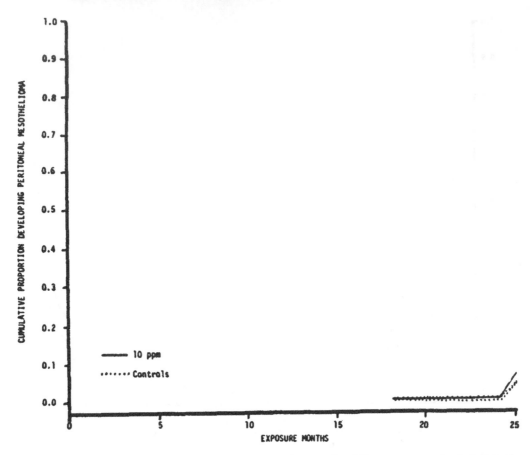

FIGURE 4.   The cumulative proportion of male rats at the control and 10 ppm exposure levels that developed peritoneal mesothelioma in the BRRC EO inhalation study.

four dose levels (0, 10, 33, and 100 ppm) had the same survival curves was always rejected at the 5% significance level. However, the null hypotheses that the survival curves at 0 and 10 ppm were equal was never rejected even at the 30% significance level. Thus, these tests imply that the survival curves at 0 and 10 ppm are similar, but that the survival curves at 33 and 100 ppm are different from that at 0 and 10 ppm. These results are consistent with the implications of Table 2 where the mean survival times without a specified response are nearly identical at 0 and 10 ppm but smaller at 33 and 100 ppm.

If a potentially fatal response is being considered, then the hazard assessment should not focus on the probability of death. After all, the probability of death is 1.0 for all animals and every person. The important thing is when that death will occur. Figure 7 indicates the cumulative proportions of male and female rats dying from any cause plotted against the number of exposure months for all four exposure levels. Figure 8 highlights the strong similarity between the mortality rates for the rats in the 10 ppm exposure group and the rats in the two control groups for each sex. The strong similarity holds for both sexes.

Tables 4 and 5 also incorporate all causes of death and summarize the changes, if any, in mortality with the level of EO vapor exposure. Table 4 indicates the rats' mean number of months without dying in the 25 month experiment for each of the four experimental dose levels. Table 5 is analogous to Table 4 but is based on only those rats which were alive at the beginning of month 17. (Tables 4 and 5 both indicate the similarity between the mortality rates at 0 and 10 ppm with a slight increase in mortality

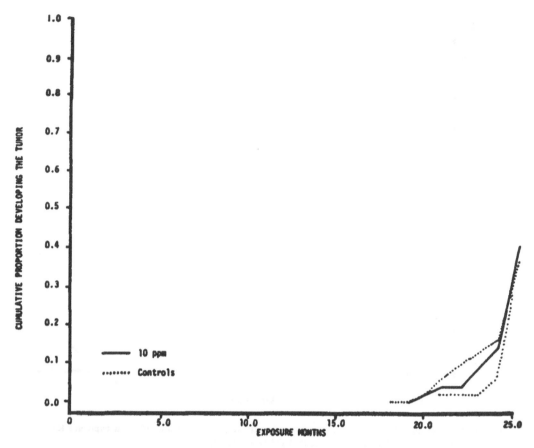

FIGURE 5. The cumulative proportion of male rats at the control and 10 ppm exposure levels that developed mononuclear cell leukemia in the BRRC EO inhalation study.

(decrease in the mean number of months without dying) at 33 ppm and the larger change at 100 ppm.

If the time to death information in the BRRC final report is summarized as in Table 6, then the null hypothesis of equal survival distributions at different exposure levels can be tested using a standard chi-square test for homogeneity. The null hypothesis that the survival curves at 0, 10, 33, and 100 ppm are all equal is rejected at the 1% significance level for both male and female rats. However, the null hypothesis that the survival curves at 0 and 10 ppm are equal is not rejected at even a 50% significance level for either male rats or female rats. These test results are consistent with the implications of Tables 4 and 5. The survival patterns and mortality rates are very similar at 0 and 10 ppm but change at the higher dose levels (33 and 100 ppm).

Some people mistakenly view the recent advances in time to response analyses as only a way to adjust simple count data for the differential mortality caused by interim sacrifices and the deaths of some test animals before the end of the experiment. Instead, time is really the key element in risk assessment and should be treated as such. Since people and experimental animals are at risk for a period of time and not just at one time, the usual risk assessments which focus on only one point in time are inadequately characterizing the real risk and may lead to false impressions. The whole time interval should be considered and not just a particular point in time like the end of the 25th month. The preceding paragraphs indicate some straightforward ways to characterize risk in a manner which reflects the time of the response. These characterizations have documented the high degree of similarity between the real risk at 0 and 10 ppm in rats.

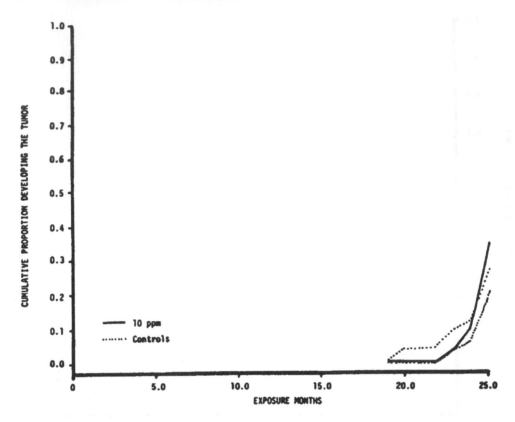

FIGURE 6.   The cumulative proportion of female rats at the control and 10 ppm exposure levels that developed mononuclear cell leukemia in the BRRC EO inhalation study.

II. SUPPORT FOR THE SIMILARITY BETWEEN RESPONSES AT 0 AND 10 PPM FROM QUANTAL RESPONSE ANALYSES

The similarity between the responses at 0 and 10 ppm is supported by several quantal response analyses as well as by the time to response analyses considered in the preceding section.

Figures 9 to 13 and Table 7 indicate the overlap between the two-sided 95% confidence intervals for the response rate at 10 ppm and the corresponding two-sided 95% confidence intervals for the response rate at 0 ppm (whether the latter are computed for the concurrent BRRC controls, the NIOSH controls, the combined BRRC and NIOSH controls, the BRRC historical controls, or all historical controls). In none of these cases are the response rates at 0 and 10 ppm significantly different at the 5% significance level.

In cases, such as EO inhalation, where high exposures are undoubtedly carcinogenic but the region of occupational or environmental concern is the relatively low dose levels, trend tests are not appropriate. Trend tests are primarily sensitive to changes in the response behaviors at the control level and the low to moderate dose levels. To illustrate this inappropriateness of trend tests, the behavior of the Cochran-Armitage trend test (see Armitage 1955) was simulated for an experimental design like that used in the BRRC study — namely, 240 animals at 0 ppm and 120 animals at each of 10, 33, and 100 ppm. The true response rates were 0, 0, 0, and 20% at 0, 10, 33 and 100 ppm, respectively. In 100 simulations of this animal bioassay the null hypothesis Ho: NO TREND was rejected every time. Clearly, rejecting "NO TREND" does not imply that a linear trend or a trend over the low doses exists.

Table 1

THE NUMBER OF MONTHS IN THE BRRC EO STUDY ON RATS UNTIL PRESCRIBED PERCENTAGES OF RATS DEVELOPED PARTICULAR RESPONSES

Dose	Prescribed percentage of rats developing the particular response			
	1%	5%	10%	20%

Response = Mononuclear Cell Leukemia in Male Rats

Dose	1%	5%	10%	20%
Controls	18	21	24	25
10 ppm	20	22	24	25
33 ppm	13	20	22	25
100 ppm	19	21	23	24

Response = Mononuclear Cell Leukemia in Female Rats

Dose	1%	5%	10%	20%
Controls	18	23	24	25
10 ppm	19	24	24	25
33 ppm	18	20	23	24
100 ppm	18	21	22	24

Response = Peritoneal Mesothelioma in Male Rats

Dose	1%	5%	10%	20%
Controls	18	25	> 25*	> 25
10 ppm	20	25	> 25	> 25
33 ppm	18	25	25	25
100 ppm	15	18	21	24

* An entry of 25 months implies that the percentage response was not reached by the end of the experiment.

Table 2

THE ESTIMATED MEAN NUMBER O MONTHS WITHOUT A RESPONSE AFTER EXPOSURE TO EO DURING THE 25 MONTH BRRC STUDY

	Mean number of months without a response	Mean percentage of experimental period without a response

Response = Mononuclear Cell Leukemia in Male Rats

	Mean number of months without a response	Mean percentage of experimental period without a response
Controls	24.25	97.0
10 ppm	24.25	97.0
33 ppm	23.76	95.0
100 ppm	23.81	95.2

Response = Mononuclear Cell Leukemia in Female Rats

	Mean number of months without a response	Mean percentage of experimental period without a response
Controls	24.48	97.9
10 ppm	24.44	97.8
33 ppm	23.89	95.6
100 ppm	23.43	93.7

Table 2 (continued)
THE ESTIMATED MEAN NUMBER O MONTHS WITHOUT A RESPONSE AFTER EXPOSURE TO EO DURING THE 25 MONTH BRRC STUDY

	Mean number of months without a response	Mean percentage of experimental period without a response
Response = Peritoneal Mesothelioma in Male Rats		
Controls	24.88	99.5
10 ppm	24.87	99.5
33 ppm	24.65	98.6
100 ppm	23.66	94.6

Table 3
THE TIME TO RESPONSE INFORMATION USED TO TEST THE EQUALITY OF THE TIME-TO-RESPONSE DISTRIBUTIONS*

Number of exposure months dose level (ppm)	Number of rats dying with the specified response				Number of rats alive at the beginning of the exposure period			
	0	10	33	100	0	10	33	100
Response = Mononuclear Cell Leukemia in Male Rats								
12—18	2	0	3	0	197	98	96	109
18—22	8	4	5	4	147	76	69	69
22—23	3	2	2	7	132	67	54	56
23—24	5	4	1	3	128	64	49	43
24—25	20	11	14	12	116	58	46	38
Response = Mononuclear Cell Leukemia in Female Rats								
12—18	1	0	1	1	198	99	98	97
18—22	4	1	4	5	146	73	69	57
22—23	5	2	2	1	132	67	58	43
23—24	4	4	9	12	122	59	52	37
24—25	8	7	8	9	96	44	42	20
Response = Peritoneal Mesothelioma in Male Rats								
12—18	1	0	1	4	197	98	96	98
18—22	1	1	1	5	147	76	69	69
22—23	0	0	0	4	132	67	54	56
23—24	0	0	0	2	128	64	49	43
24—25	2	2	5	7	116	58	46	38

* None of the three responses considered herein ever occurred before 12 months of exposure.

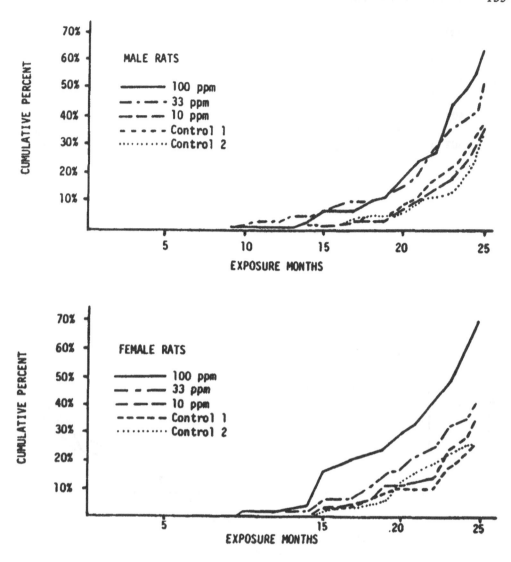

FIGURE 7. Cumulative percentages of rats dying or sacrificed in a moribund condition at the control and all exposure levels in the BRRC EO inhalation study.

Quantal response models (multistage, Weibull, etc.) are based on very simplified interpretations of cancer and do not even attempt to be detailed biological models of the carcinogenic process. Furthermore, the quantal response models do not provide sufficiently flexible models here to reliably reflect either the similarity between the response rates at 0 to 10 ppm or the observed behaviors at 33 and 100 ppm. For example, Figures 14 and 15 indicate the fitted multistage and Weibull models for the observed quantal response data when the specified brain tumor responses are malignant reticulosis or glial cell tumors (Figure 14) and malignant reticulosis, glial cell tumors, or granular cell tumors (Figure 15). The low dose behavior of the fitted models is almost the same in Figure 14 as it is in Figure 15 despite the fact that the observed response proportions decrease from 0 to 10 ppm in Figure 14 and increase from 0 to 10 ppm in Figure 15. Furthermore, in both figures the fitted models do not pass close to the observed response proportions at either 33 or 100 ppm but rather try to "compromise" by passing through the middle of the relatively flat dose-response curves at high doses. This "compromising" in the fitted models can usually be alleviated by fitting the models to the observed proportions excluding the highest dose level(s).

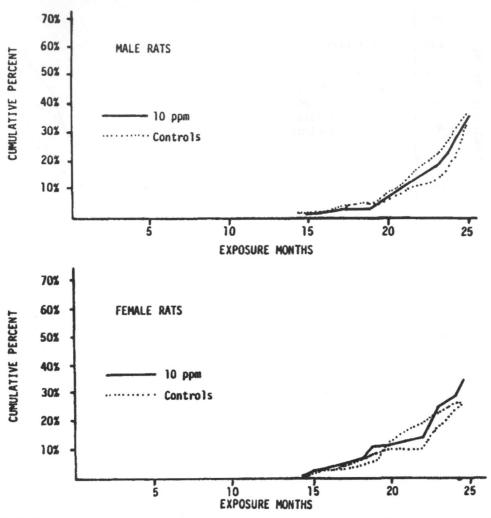

FIGURE 8.   Cumulative percentages of rats dying or sacrificed in a moribund condition at the control and 10 ppm exposure levels in the BRRC EO inhalation study.

Table 4
THE ESTIMATED MEAN NUMBER OF MONTHS
WITHOUT DYING AFTER EXPOSURE TO EO
DURING THE 25 MONTH BRRC STUDY

	Dose level			
	Control	10	33	100
Male rats				
Mean number of months without dying	23.56	23.65	22.38	22.19
Mean percentage of experimental period survived	94.2	94.6	89.5	88.8
Female rats				
Mean number of months without dying	23.62	23.44	22.70	20.93
Mean percentage of experimental period survived	94.5	93.8	90.8	83.7

Table 5

THE ESTIMATED MEAN NUMBER OF MONTHS
WITHOUT DYING AFTER EXPOSURE TO EO
DURING THE 25 MONTH BRRC STUDY AMONG
RATS SURVIVING TO THE BEGINNING OF
MONTH 17

	Dose level			
	Control	10	33	100
Male rats				
Mean number of months without dying	23.79	23.88	23.37	22.86
Mean percentage of experimental period survived	95.2	95.5	93.5	91.5
Female rats				
Mean number of months without dying	24.02	23.73	23.35	22.58
Mean percentage of experimental period survived	96.1	94.9	93.4	90.3

Table 6

THE OBSERVED TIMES TO DEATH WITHOUT
REGARD FOR THE CAUSE OF DEATH IN THE BRRC
STUDY OF EO INHALATION BY RATS[*]

	Exposure levels (ppm)							
	Male rats				Female rats			
Time of death (months)	0	10	33	100	0	10	33	100
0—12	1	1	3	1	1	0	1	2
12—15	3	0	4	6	5	2	5	14
15—18	6	2	3	3	4	4	4	6
18—21	9	6	8	11	12	5	9	9
21—22	6	3	7	2	2	1	2	5
22—23	4	3	5	13	10	8	6	6
23—24	12	6	3	5	5	3	2	9
24—25	19	7	7	8	1	2	2	2
>25	97	51	39	30	95	42	40	18

[*] All interim sacrifice rats are omitted.

Since quantal response modeling is primarily curve fitting as opposed to biological modeling, it is important to consider not only the associated fitted models and upper bounds on risk but also the lower bounds on risk. The same statistical procedures used to generate upper bounds (upper confidence limits) can be used to generate lower bounds (lower confidence limits). Not only do these lower bounds provide information by themselves, but also the difference between the upper and lower bounds can be informative. This difference provides an indication of how specific or vague the bounds really are. Also, the farther apart the upper and lower bounds are using the same procedures, the less likely the true value is to being near either bound.

The fitted quantal response models and their associated upper and lower bounds are described in Tables 8 to 15 and graphed in Figures 16 to 23. These values have been

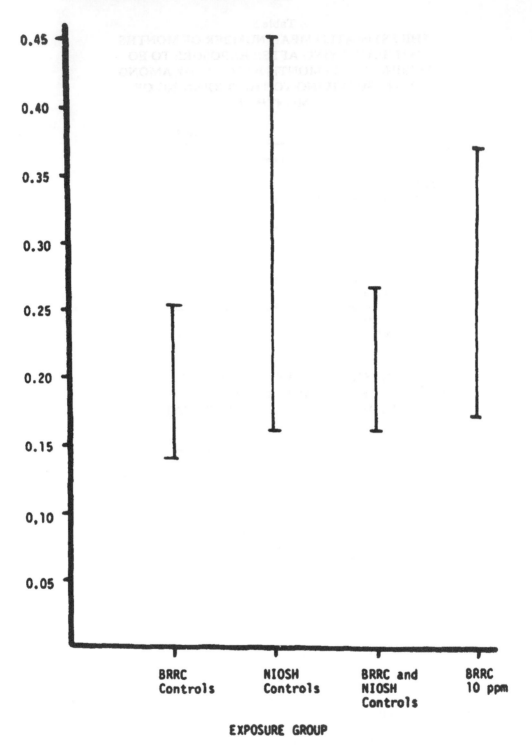

FIGURE 9.   Comparisons which are independent of the quantal response modeling: 95% two-sided confidence intervals on the probability of a male rat developing a mononuclear cell leukemia at the control and 10 ppm exposure levels.

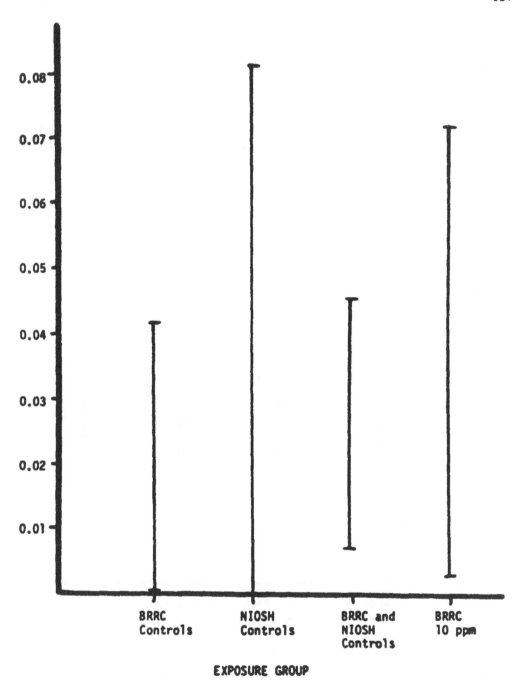

FIGURE 10. Comparisons which are independent of the quantal response modeling: 95% two-sided confidence intervals on the probability of a male rat developing a peritoneal mesothelioma at the control and 10 ppm exposure levels.

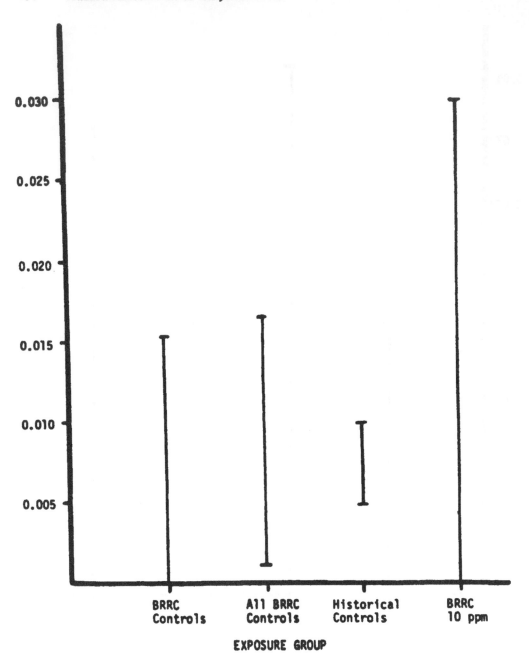

FIGURE 11.   Comparisons which are independent of the quantal response modeling: 95% two-sided confidence intervals on the probability of a male rat developing a brain glioma at the control and 10 ppm exposure levels.

computed from the BRRC EO inhalation quantal response data on rats for the following responses:

1.   Brain gliomas in male rats
2.   Peritoneal mesothelioma in male rats
3.   Brain gliomas in female rats
4.   Mononuclear cell leukemia in female rats

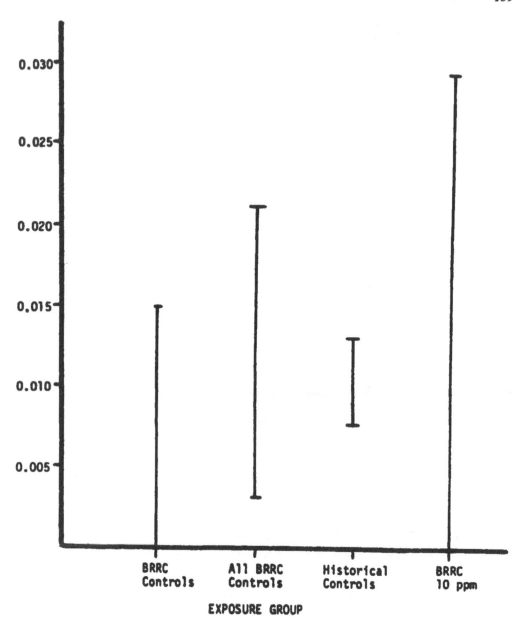

FIGURE 12. Comparisons which are independent of the quantal response modeling: 95% two-sided confidence intervals on the probability of a female rat developing a brain glioma at the control and 10 ppm exposure levels.

The analyses have been done using the quantal response data at 0, 10, and 33 ppm. The data at 100 ppm has been excluded since the models do not seem to fit well the usually relatively flat dose-response relationship between 33 and 100 ppm. Tables 8 to 11 show the experimental data, fitted multistage model estimates, 95% upper confidence limits, and 95% lower confidence limits for the responses 1 to 4, respectively. Figures 16 to 19 indicate plots of the fitted multistage model estimates, 95% upper confidence limits, and 95% lower confidence limits over the 0 to 10 ppm range of exposure levels for the responses (1 to 4), respectively. The empirical behavior in these tables and figures is not unique to the multistage model. Tables 12 to 15 and Figures 20 to 23 are the analogous tables and figures for the Weibull model.

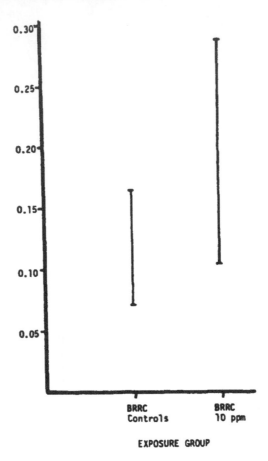

FIGURE 13.   Comparisons which are independent of the quantal response modeling: 95% two-sided confidence intervals on the probability of a female rat developing a mononuclear cell leukemia at the control and 10 ppm exposure levels.

When both the upper and lower bounds are generated for the quantal response models and examined, it again becomes apparent that the risks in the 0 to 10 ppm exposure range are *not* significantly different.

III. ALTERNATIVE RISK CHARACTERIZATIONS BASED ON TIME TO RESPONSE

A practical way to characterize the effect of a particular exposure is to describe the corresponding average amount of time in a specified observation period during which the subject is free from a specified response. This characterization has been called the mean free period. For example, if the observation period is 25 months as it was in the BRRC study on EO vapor inhalation, then the maximum mean free period is 25 months. Furthermore, if the specified response is death with mononuclear cell leukemia in female rats, then a mean free period of 24.0 months would imply that the mean number of months that a female rat would be observed without that rat dying with mononuclear cell leukemia would be 24.0 months.

Table 2 indicates the observed mean free periods in the BRRC study for 3 specified responses (death of a male rat with mononuclear cell leukemia, death of a male rat with peritoneal mesothelioma, and death of a female rat with mononuclear cell leukemia). Figure 24 is a graphical presentation of the values in Table 2. (Since the observed

Table 7
COMPARISONS WHICH ARE INDEPENDENT OF THE QUANTAL RESPONSE MODELING: 95% TWO-SIDED CONFIDENCE INTERVALS ON THE PROBABILITY OF A RAT DEVELOPING A SPECIFIED RESPONSE AT THE CONTROL AND 10 PPM EXPOSURE LEVELS

Animal	Specified response	Exposure group	Observed number of animals with the specified response	Number of animals at risk	95% Two-sided confidence interval	
					Lower bound on probability	Upper bound on probability
Male rat	Mononuclear cell leukemia	BRRC controls	38	193	0.1408	0.2530
Male rat	Mononuclear cell leukemia	NIOSH controls	12	39	0.1628	0.4525
Male rat	Mononuclear cell leukemia	BRRC and NIOSH controls	50	232	0.1626	0.2684
Male rat	Mononuclear cell leukemia	BRRC 10 ppm	21	77	0.1733	0.3722
Male rat	Peritoneal mesothelioma	BRRC controls	4	187	0.0007	0.0421
Male rat	Peritoneal mesothelioma	NIOSH controls	3	78	0.0000	0.0811
Male rat	Peritoneal mesothelioma	BRRC and NIOSH controls	7	265	0.0071	0.0457
Male rat	Peritoneal mesothelioma	BRRC 10 ppm	3	88	0.0000	0.0720
Male rat	Brain glioma	BRRC controls	1	196	0.0000	0.0151
Male rat	Brain glioma	All BRRC controls	7	571	0.0032	0.0213
Male rat	Brain glioma	Historical controls	57	5450	0.0078	0.0132
Male rat	Brain glioma	BRRC 10 ppm	1	99	0.0000	0.0298
Female rat	Brain glioma	BRRC controls	1	194	0.0000	0.0152
Female rat	Brain glioma	All BRRC controls	5	567	0.0011	0.0165
Female rat	Brain glioma	Historical controls	43	5524	0.0055	0.0101
Female rat	Brain glioma	BRRC 10 ppm	1	98	0.0000	0.0301
Female rat	Mononuclear cell leukemia	BRRC controls	22	186	0.0719	0.1647
Female rat	Mononuclear cell leukemia	BRRC 10 ppm	14	71	0.1046	0.2897

mean free periods at 0, 10, 33, and 100 ppm were relatively large fractions of the 25 month experimental period, the vertical axis in Figure 24 has been expanded. Also in Figure 24 the observed mean free periods at the experimental dose levels have been connected by straight lines for ease of interpretation.)

By combining all causes of death, the observed mean free period can be determined when the specified response is simply death. Such a mean free period represents the "total" risk of a particular exposure. Table 4 indicates the observed mean free periods in the BRRC study for two specified responses (death of a male rat and death of a female rat). Figure 25 shows a piecewise linear plot of these observed mean free periods.

The rate of decrease in the observed mean free periods in Figure 25 is greater for the

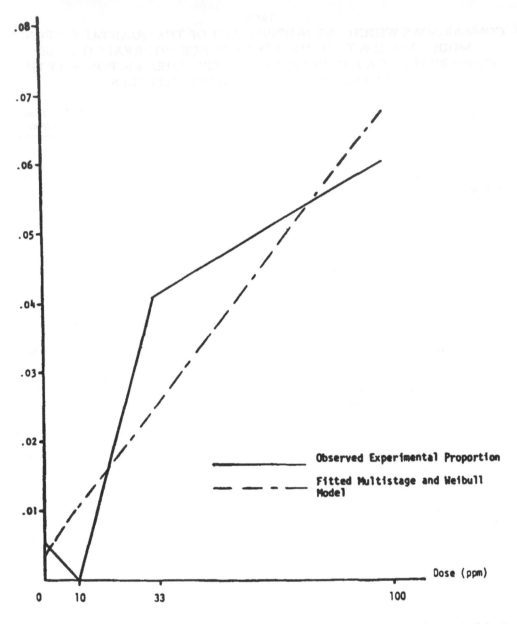

FIGURE 14.   The observed and fitted proportions of male rats with malignant reticulosis and glial cell tumors based on 18 and 24 month sacrifice and dead/euthanatized moribund male rats in the BRRC EO inhalation study.

female rats than the male rats. The rate of change in the observed mean free period is

$$\frac{23.44 - 23.62}{10 - 0} = -0.018$$

between 0 and 10 ppm. If a linear interpolation between 0 and 10 ppm holds, this decrease implies that a female rat's lifespan during a 25 month experiment decreases approximately 0.018 months (approximately 0.54 days, 13 hr, or 0.072%) with each 1 ppm increase in exposure from 0 to 10 ppm. Some implications of this rate of decrease in the mean free period are shown in Table 16.

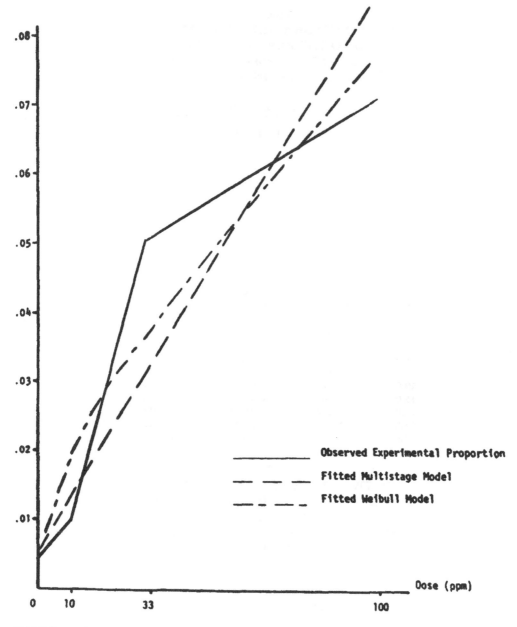

FIGURE 15. The observed and fitted proportions of male rats with malignant reticulosis, glial cell tumors, or granular cell tumors based on 18 and 24 month sacrifice and dead/euthanatized moribund male rats in the BRRC EO inhalation study.

The preceding description of the mean free periods utilized linear interpolation between the observed mean free periods at 0, 10, 33, and 100 ppm. Alternatively, the mean free periods can be estimated by fitting a time to response model to the experimental time to response data and then computing the estimated mean free period from the estimated model. The Hartley-Sielken time to response model,

$$P(T;d) = \text{Probability that a rat will develop the specified response by time } T \text{ at dose level } d$$

$$= 1.0 - \exp\left\{-\left[\sum_{i=0}^{3} \alpha_i d^i\right]\left[\sum_{j=1}^{4} \beta_j (T - LP)^j\right]\right\}$$

Table 8

THE TWO-STAGE MULTISTAGE MODEL APPLIED TO THE BRRC EO INHALATION STUDY QUANTAL RESPONSE DATA ON BRAIN GLIOMAS IN MALE RATS

Dose: response frequency		
0 ppm: 1/196	10 ppm: 1/99	33 ppm: 5/98

Characteristics of multistage modeling		

Dose (ppm)	95% Lower confidence limit on added risk	Fitted model estimate of added risk	95% Upper confidence limit on added risk
0.5	−0.0011	0.00006	0.0012
1.0	−0.0022	0.00014	0.0024
2.0	−0.0041	0.00036	0.0049
3.0	−0.0058	0.00066	0.0075
4.0	−0.0073	0.00105	0.0101
5.0	−0.0086	0.00151	0.0128
6.0	−0.0097	0.00205	0.0156
7.0	−0.0105	0.00268	0.0185
8.0	−0.0111	0.00338	0.0214
9.0	−0.0115	0.00416	0.0240
10.0	−0.0117	0.00502	0.0275
15.0	−0.0092	0.01053	0.0438
20.0	−0.0012	0.01798	0.0618
25.0	0.0122	0.02734	0.0812
33.0	0.0443	0.04615	0.1153

Table 9

THE TWO-STAGE MULTISTAGE MODEL APPLIED TO THE BRRC EO INHALATION STUDY QUANTAL RESPONSE DATA ON PERITONEAL MESOTHELIOMA IN MALE RATS

Dose: response frequency		
0 ppm: 4/187	10 ppm: 3/88	33 ppm: 7/82

Characteristics of multistage modeling		

Dose (ppm)	95% Lower confidence limit on added risk	Fitted model estimate of added risk	95% Upper confidence limit on added risk
0.5	−0.0016	0.0005	0.0019
1.0	−0.0032	0.0010	0.0038
2.0	−0.0061	0.0020	0.0077
3.0	−0.0087	0.0032	0.0116
4.0	−0.0110	0.0043	0.0155
5.0	−0.0130	0.0056	0.0195
6.0	−0.0146	0.0069	0.0236
7.0	−0.0160	0.0083	0.0277

Table 9 (continued)
THE TWO-STAGE MULTISTAGE MODEL APPLIED TO THE BRRC EO INHALATION STUDY QUANTAL RESPONSE DATA ON PERITONEAL MESOTHELIOMA IN MALE RATS

Dose: response frequency

0 ppm: 4/187	10 ppm: 3/88	33 ppm: 7/82

Characteristics of multistage modeling

Dose (ppm)	95% Lower confidence limit on added risk	Fitted model estimate of added risk	95% Upper confidence limit on added risk
8.0	−0.0170	0.0097	0.0318
9.0	−0.0177	0.0112	0.0360
10.0	−0.0180	0.0127	0.0402
15.0	−0.0150	0.0213	0.0620
20.0	−0.0040	0.0314	0.0847
25.0	0.0147	0.0428	0.1084
33.0	0.0596	0.0640	0.1478

Table 10
THE TWO-STAGE MULTISTAGE MODEL APPLIED TO THE BRRC EO INHALATION STUDY QUANTAL RESPONSE DATA ON BRAIN GLIOMAS IN FEMALE RATS

Dose: response frequency

0 ppm: 1/194	10 ppm: 1/98	33 ppm: 3/99

Characteristics of multistage modeling

Dose (ppm)	95% Lower confidence limit on added risk	Fitted model estimate of added risk	95% Upper confidence limit on added risk
0.5	−0.0010	0.0002	0.0009
1.0	−0.0019	0.0004	0.0018
2.0	−0.0036	0.0008	0.0036
3.0	−0.0052	0.0013	0.0054
4.0	−0.0067	0.0017	0.0073
5.0	−0.0080	0.0022	0.0091
6.0	−0.0091	0.0028	0.0110
7.0	−0.0101	0.0033	0.0129
8.0	−0.0109	0.0039	0.0148
9.0	−0.0116	0.0044	0.0168
10.0	−0.0122	0.0050	0.0187
15.0	−0.0126	0.0084	0.0288
20.0	−0.0091	0.0124	0.0393
25.0	−0.0019	0.0168	0.0503
33.0	0.0144	0.0251	0.0687

Table 11

THE TWO-STAGE MULTISTAGE MODEL APPLIED TO THE BRRC EO INHALATION STUDY QUANTAL RESPONSE DATA ON MONONUCLEAR CELL LEUKEMIA IN FEMALE RATS

	Dose: response frequency		
	0 ppm: 22/186	10 ppm: 14/71	33 ppm: 24/72

	Characteristics of multistage modeling		
Dose (ppm)	95% Lower confidence limit on added risk	Fitted model estimate of added risk	95% Upper confidence limit on added risk
0.5	−0.0016	0.0038	0.0058
1.0	−0.0031	0.0075	0.0115
2.0	−0.0056	0.0150	0.0228
3.0	−0.0075	0.0224	0.0340
4.0	−0.0087	0.0297	0.0450
5.0	−0.0092	0.0370	0.0559
6.0	−0.0091	0.0442	0.0667
7.0	−0.0084	0.0513	0.0773
8.0	−0.0070	0.0584	0.0878
9.0	−0.0049	0.0654	0.0981
10.0	−0.0022	0.0724	0.1083
15.0	0.0205	0.1063	0.1574
20.0	0.0579	0.1388	0.2033
25.0	0.1078	0.1699	0.2464
33.0	0.2071	0.2171	0.3096

Table 12

THE WEIBULL MODEL APPLIED TO THE BRRC EO INHALATION STUDY QUANTAL RESPONSE DATA ON BRAIN GLIOMAS IN MALE RATS

	Dose: response frequency		
	0 ppm: 1/196	10 ppm: 1/99	33 ppm: 5/98

	Characteristics of Weibull modeling		
Dose (ppm)	95% Lower confidence limit on added risk	Fitted model estimate of added risk	95% Upper confidence limit on added risk
0.5	−0.00022	0.00002	0.00026
1.0	−0.00066	0.00007	0.00079
2.0	−0.00188	0.00025	0.00237
3.0	−0.00337	0.00052	0.00442
4.0	−0.00497	0.00090	0.00677
5.0	−0.00661	0.00137	0.00934
6.0	−0.00822	0.00192	0.01206
7.0	−0.00975	0.00236	0.01488
8.0	−0.01117	0.00329	0.01775

Table 12 (continued)
THE WEIBULL MODEL APPLIED TO THE BRRC EO INHALATION STUDY QUANTAL RESPONSE DATA ON BRAIN GLIOMAS IN MALE RATS

Dose: response frequency

0 ppm: 1/196	10 ppm: 1/99	33 ppm: 5/98

Characteristics of Weibull modeling

Dose (ppm)	95% Lower confidence limit on added risk	Fitted model estimate of added risk	95% Upper confidence limit on added risk
9.0	−0.01244	0.00411	0.02065
10.0	−0.01353	0.00500	0.02353
15.0	−0.01584	0.01066	0.03716
20.0	−0.01208	0.01821	0.04851
25.0	−0.00308	0.02755	0.05817
33.0	0.00840	0.04592	0.08343

Table 13
THE WEIBULL MODEL APPLIED TO THE BRRC EO INHALATION STUDY QUANTAL RESPONSE DATA ON PERITONEAL MESOTHELIOMA IN MALE RATS

Dose: response frequency

0 ppm: 4/187	10 ppm: 3/88	33 ppm: 7/82

Characteristics of Weibull modeling

Dose (ppm)	95% Lower confidence limit on added risk	Fitted model estimate of added risk	95% Upper confidence limit on added risk
0.5	−0.0018	0.0002	0.0023
1.0	−0.0039	0.0005	0.0050
2.0	−0.0079	0.0014	0.0107
3.0	−0.0114	0.0024	0.0163
4.0	−0.0145	0.0036	0.0217
5.0	−0.0171	0.0049	0.0269
6.0	−0.0192	0.0063	0.0319
7.0	−0.0209	0.0078	0.0365
8.0	−0.0222	0.0094	0.0409
9.0	−0.0231	0.0110	0.0451
10.0	−0.0236	0.0127	0.0490
15.0	−0.0208	0.0221	0.0650
20.0	−0.0116	0.0326	0.0769
25.0	0.0004	0.0441	0.0879
33.0	0.0103	0.0640	0.1176

Table 14

THE WEIBULL MODEL APPLIED TO THE BRRC EO INHALATION STUDY QUANTAL RESPONSE DATA ON BRAIN GLIOMAS IN FEMALE RATS

Dose: response frequency

| 0 ppm: 1/194 | 10 ppm: 1/98 | 33 ppm: 3/99 |

Characteristics of Weibull modeling

Dose (ppm)	95% Lower confidence limit on added risk	Fitted model estimate of added risk	95% Upper confidence limit on added risk
0.5	−0.00105	0.00009	0.00123
1.0	−0.00221	0.00022	0.00266
2.0	−0.00439	0.00057	0.00554
3.0	−0.00635	0.00099	0.00834
4.0	−0.00807	0.00146	0.01100
5.0	−0.00954	0.00198	0.01350
6.0	−0.01079	0.00253	0.01585
7.0	−0.01181	0.00311	0.01805
8.0	−0.01263	0.00374	0.02010
9.0	−0.01325	0.00438	0.02200
10.0	−0.01367	0.00505	0.02377
15.0	−0.01333	0.00873	0.03078
20.0	−0.00995	0.01285	0.03565
25.0	−0.00564	0.01734	0.04032
33.0	−0.00443	0.02515	0.05473

Table 15

THE WEIBULL MODEL APPLIED TO THE BRRC EO INHALATION STUDY QUANTAL RESPONSE DATA ON MONONUCLEAR CELL LEUKEMIA IN FEMALE RATS

Dose: response frequency

| 0 ppm: 22/186 | 10 ppm: 14/71 | 33 ppm: 24/72 |

Characteristics of Weibull modeling

Dose (ppm)	95% Lower confidence limit on added risk	Fitted model estimate of added risk	95% Upper confidence limit on added risk
0.5	−0.0161	0.0053	0.0267
1.0	−0.0234	0.0100	0.0434
2.0	−0.0310	0.0188	0.0685
3.0	−0.0339	0.0270	0.0880
4.0	−0.0340	0.0350	0.1040
5.0	−0.0322	0.0428	0.1177
6.0	−0.0291	0.0503	0.1296
7.0	−0.0248	0.0577	0.1402
8.0	−0.0198	0.0649	0.1496

Table 15 (continued)
THE WEIBULL MODEL APPLIED TO THE BRRC EO INHALATION STUDY QUANTAL RESPONSE DATA ON MONONUCLEAR CELL LEUKEMIA IN FEMALE RATS

Dose: response frequency

| 0 ppm: 22/186 | 10 ppm: 14/71 | 33 ppm: 24/72 |

Characteristics of Weibull modeling

Dose (ppm)	95% Lower confidence limit on added risk	Fitted model estimate of added risk	95% Upper confidence limit on added risk
9.0	−0.0142	0.0720	0.1581
10.0	−0.0080	0.0789	0.1658
15.0	0.0269	0.1120	0.1971
20.0	0.0621	0.1429	0.2237
25.0	0.0908	0.1719	0.2530
33.0	0.1157	0.2151	0.3144

where $T - LP$ is the number of months beyond an assumed LP month latency period ($LP = 12$ for carcinogenic responses since 13 months was the earliest response time observed and $LP = 0$ for the "time to death from any cause" data), was fit to the time to response information in Tables 17 to 21 which follow from the BRRC final report (the rats "lost to follow-up" are excluded from the time to carcinogenic response analyses). Tables 22 and 23 and Figures 26 and 27 indicate the corresponding estimated mean free periods.

In Figure 27 the estimated mean free period when the specified response is death from any cause decreases more rapidly for female rats than it does for male rats. Some implications of this rate of decrease are shown in Table 24.

The implications in Figures 25 and 27 are nearly the same. Hence, regardless of whether a crude interpolation or a more sophisticated time to response model estimate is used, the practical implications about the mean free period (the average length of time without dying) are nearly the same. The mean free period does not decrease appreciably in the low dose region.

This section has illustrated an alternative characterization of the real risk of an exposure. This risk characterization is based on the mean free period (the mean amount of time without a specified response) and focuses on not just the presence or absence of a response but also when the specified response may occur.

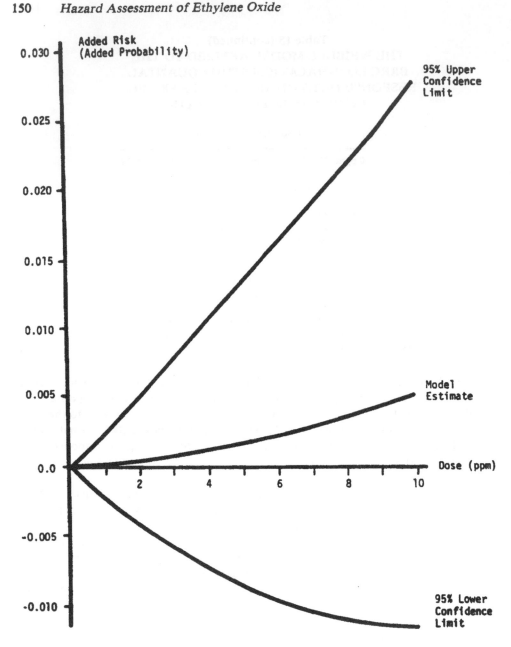

FIGURE 16.   The two-stage multistage model applied to the BRRC EO inhalation study quantal response data on brain gliomas in male rats.

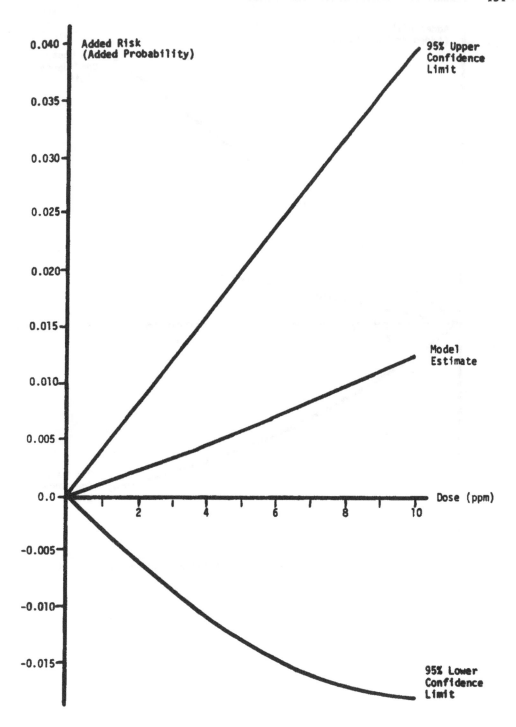

FIGURE 17. The two-stage multistage model applied to the BRRC EO inhalation study quantal response data on peritoneal mesothelioma in male rats.

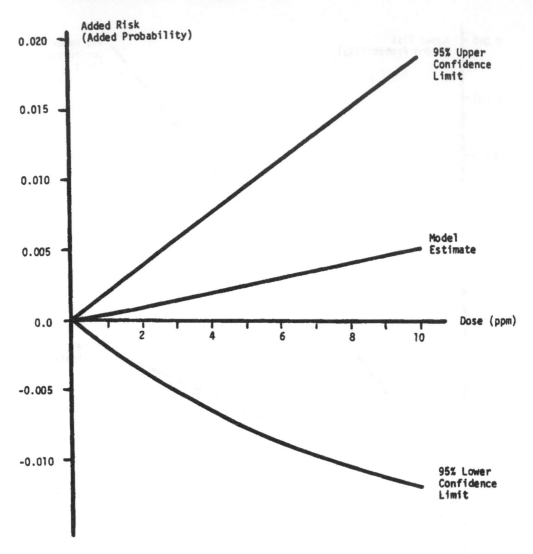

FIGURE 18.   The two-stage multistage model applied to the BRRC EO inhalation study quantal response data on brain gliomas in female rats.

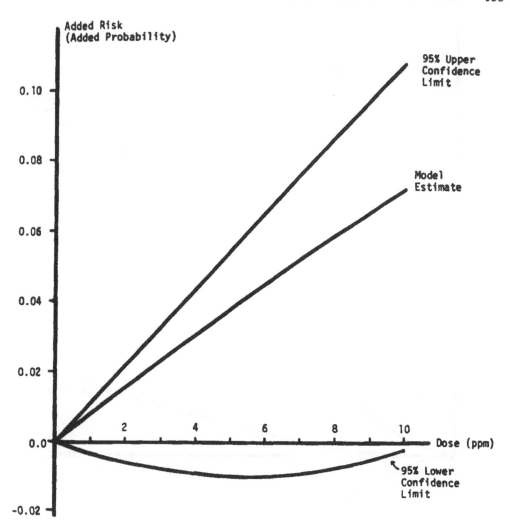

FIGURE 19. The two-stage multistage model applied to the BRRC EO inhalation study quantal response data on mononuclear cell leukemia in female rats.

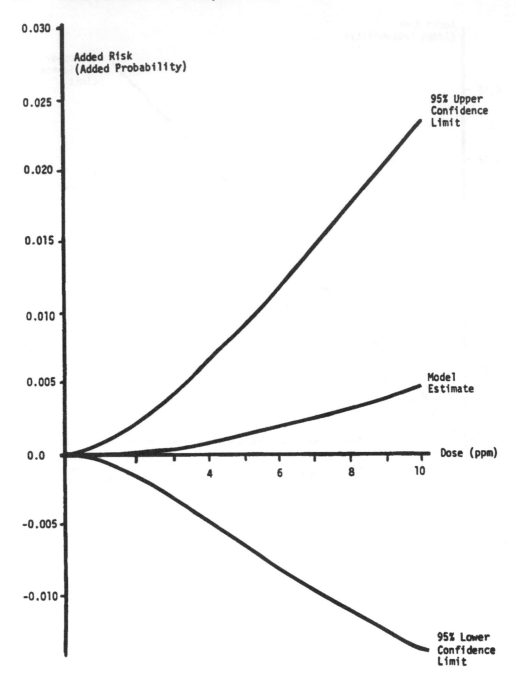

FIGURE 20.   The Weibull model applied to the BRRC EO inhalation study quantal response data on brain gliomas in male rats.

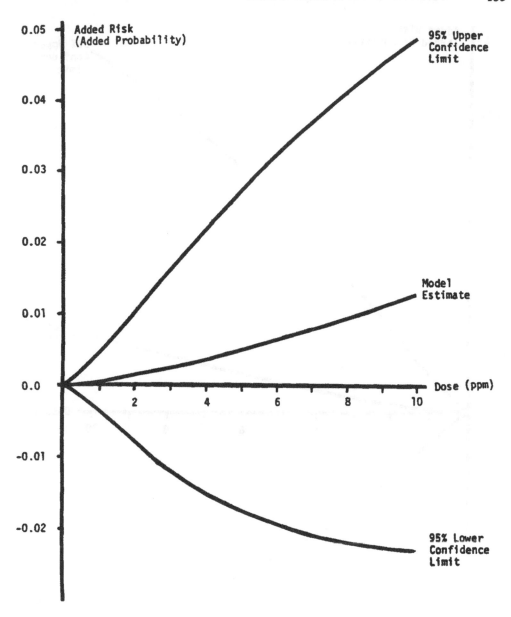

FIGURE 21. The Weibull model applied to the BRRC EO inhalation study quantal response data on peritoneal mesothelioma in male rats.

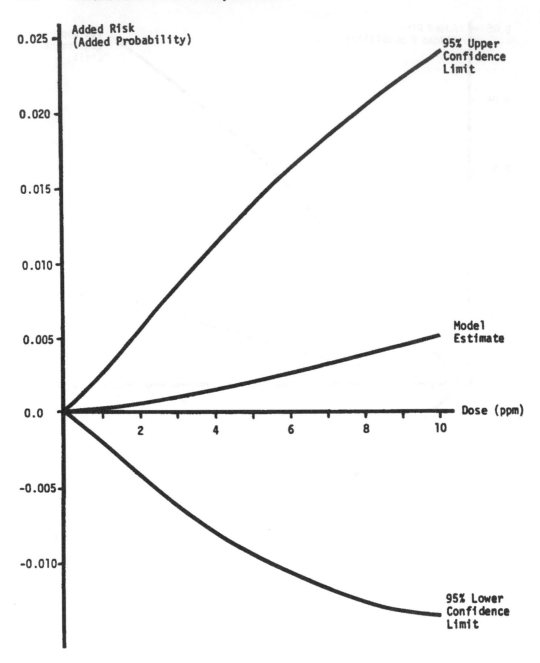

FIGURE 22.   The Weibull model applied to the BRRC EO inhalation study quantal response data on brain gliomas in female rats.

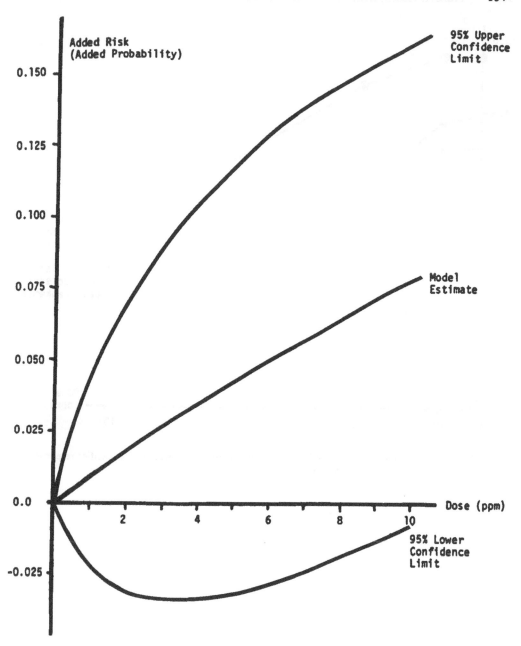

FIGURE 23. The Weibull model applied to the BRRC EO inhalation study quantal response data on mononuclear cell leukemia in female rats.

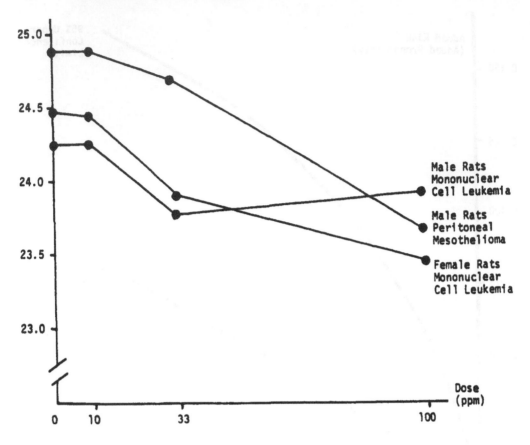

FIGURE 24.   The observed mean free periods (mean number of months without the specified response) for rats exposed to EO vapor in the BRRC study.

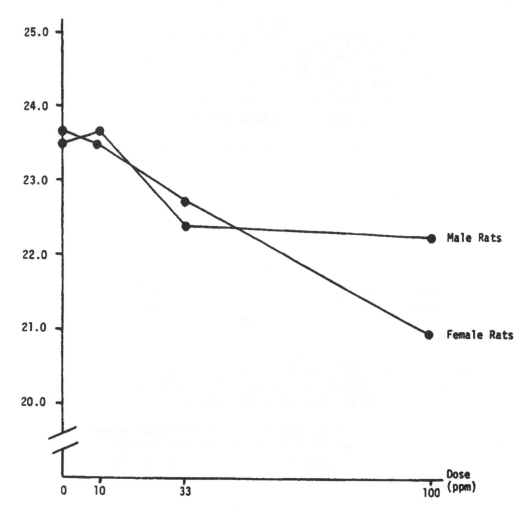

FIGURE 25. The observed mean number of months without dying (mean free period) for rats exposed to EO in the BRRC study.

Table 16
DECREASES IN THE MEAN FREE PERIOD
(AVERAGE NUMBER OF MONTHS WITHOUT
DYING) FOR FEMALE RATS IN THE 25 MONTH
BRRC EO STUDY BASED ON THE
CORRESPONDING OBSERVED MEAN FREE
PERIODS AT 0 AND 10 PPM

Dose (ppm)	Decrease in female rat's mean free period	Percentage decrease relative to 25 month lifespan	Corresponding decrease in a 70 year lifespan
0.0	0.0	0.0	0.0
1.0	0.018 months	0.072	0.60 months
5.0	0.09 months	0.360	3.02 months
10.0	0.18 months	0.720	6.05 months
1.85	1.0 day	0.133	1.12 months
0.077	1.0 hr	0.006	0.05 months
1.65	0.030 months	0.119	1.0 month
0.41	0.007 months	0.027	1.0 week
0.054	0.001 months	0.004	1.0 day
0.002	0.00004 months	0.00014	1.0 hr

Table 17
TIME TO DEATH WITH MONONUCLEAR CELL
LEUKEMIA AMONG MALE RATS EXPOSED TO EO
VAPOR IN THE BRRC STUDY

Exposure months	Number of deaths with mononuclear cell leukemia (dose level/ppm)				Number of deaths without mononuclear cell leukemia (dose level/ppm)			
	0	10	33	100	0	10	33	100
1	0	0	0	0	0	0	0	0
2	0	0	0	0	0	0	0	0
3	0	0	0	0	0	0	0	0
4	0	0	0	0	0	0	0	0
5	0	0	0	0	1	0	0	0
6	0	0	0	0	20	0	0	10
7	0	0	0	0	0	0	0	0
8	0	0	0	0	0	0	0	0
9	0	0	0	0	0	1	1	0
10	0	0	0	0	0	0	1	0
11	0	0	0	0	0	0	1	0
12	0	0	0	0	20	1	1	11
13	0	0	1	0	0	0	1	0
14	0	0	0	0	2	0	0	2
15	0	0	1	0	2	0	1	4
16	0	0	0	0	0	1	2	0
17	0	0	1	0	3	1	0	0
18	2	0	0	0	41	0	2	23
19	0	0	0	1	0	0	2	0
20	1	2	1	1	3	2	1	4
21	4	1	2	2	1	1	2	3
22	3	1	2	0	1	2	5	2
23	3	2	2	7	1	1	3	6
24	5	4	1	3	6	1	2	2
25	20	11	14	12	95	47	31	26

Table 18
TIME TO DEATH WITH PERITONEAL MESOTHELIOMA AMONG MALE RATS EXPOSED TO EO VAPOR IN THE BRRC STUDY

Exposure months	Number of deaths with peritoneal mesothelioma (dose level/ppm)				Number of deaths without peritoneal mesothelioma (dose level/ppm)			
	0	10	33	100	0	10	33	100
1	0	0	0	0	0	0	0	0
2	0	0	0	0	0	0	0	0
3	0	0	0	0	0	0	0	0
4	0	0	0	0	0	0	0	0
5	0	0	0	0	0	0	0	0
6	0	0	0	0	20	0	0	10
7	0	0	0	0	0	0	0	0
8	0	0	0	0	0	0	0	0
9	0	0	0	0	0	1	1	0
10	0	0	0	0	0	0	1	0
11	0	0	0	0	0	0	1	0
12	0	0	0	0	20	0	1	11
13	0	0	0	0	0	0	2	0
14	0	0	0	0	0	0	0	2
15	0	0	0	1	2	0	2	3
16	0	0	0	0	0	1	2	0
17	0	0	0	0	3	1	1	0
18	1	0	1	3	41	10	12	20
19	0	0	0	0	0	0	2	1
20	1	1	0	2	3	3	2	3
21	0	0	0	3	4	2	4	2
22	0	0	1	0	5	3	5	2
23	0	0	0	4	4	3	5	9
24	0	0	0	2	10	6	3	3
25	2	2	5	7	114	56	40	31

Table 19
TIME TO DEATH WITH MONONUCLEAR CELL LEUKEMIA AMONG FEMALE RATS EXPOSED TO EO VAPOR IN THE BRRC STUDY

Exposure months	Number of deaths with mononuclear cell leukemia (dose level/ppm)				Number of deaths without mononuclear cell leukemia (dose level/ppm)			
	0	10	33	100	0	10	33	100
1	0	0	0	0	0	0	0	0
2	0	0	0	0	0	0	0	0
3	0	0	0	0	0	0	0	0
4	0	0	0	0	0	0	1	0
5	0	0	0	0	0	0	0	0
6	0	0	0	0	0	0	0	0
7	0	0	0	0	20	0	0	10
8	0	0	0	0	0	0	0	0
9	0	0	0	0	0	0	0	0
10	0	0	0	0	0	0	0	2
11	0	0	0	0	0	0	0	0
12	0	0	0	0	20	1	0	10
13	0	0	0	0	0	0	1	1

Table 19 (continued)
TIME TO DEATH WITH MONONUCLEAR CELL LEUKEMIA AMONG FEMALE RATS EXPOSED TO EO VAPOR IN THE BRRC STUDY

Exposure months	Number of deaths with mononuclear cell leukemia (dose level/ppm)				Number of deaths without mononuclear cell leukemia (dose level/ppm)			
	0	10	33	100	0	10	33	100
14	0	0	0	0	3	0	0	1
15	0	0	0	0	5	2	4	11
16	0	0	0	0	1	1	0	2
17	0	0	0	0	0	2	1	3
18	1	0	1	1	42	1	2	20
19	1	1	2	0	1	3	2	1
20	2	0	1	1	4	0	0	2
21	1	0	1	1	2	1	3	2
22	0	0	0	3	2	0	2	0
23	5	2	2	1	4	4	4	4
24	4	4	9	12	22	11	1	5
25	8	7	8	9	87	37	34	11

Table 20
TIME TO DEATH AMONG MALE RATS EXPOSED TO EO VAPOR IN THE BRRC STUDY

Exposure months	Number of deaths (dose level/ppm)				Number of rats sacrificed at necropsy interval (dose level/ppm)			
	0	10	33	100	0	10	33	100
1	0	0	0	0	0	0	0	0
2	0	0	0	0	0	0	0	0
3	0	0	0	0	2	1	1	1
4	0	0	0	0	0	0	0	0
5	1	0	0	0	0	0	0	0
6	0	0	0	0	20	10	10	10
7	0	0	0	0	0	0	0	0
8	0	0	0	0	0	0	0	0
9	0	1	1	0	0	0	0	0
10	0	0	1	0	0	0	0	0
11	0	0	1	0	0	0	0	0
12	0	0	0	1	20	10	10	10
13	0	0	2	0	0	0	0	0
14	1	0	0	2	1	0	0	0
15	2	0	2	4	0	0	0	0
16	0	1	2	0	0	0	0	0
17	3	1	1	0	0	0	0	0
18	3	0	0	3	40	20	20	20
19	0	0	2	1	0	0	0	0
20	4	4	2	5	0	0	0	0
21	5	2	4	5	0	0	0	0
22	6	3	7	2	0	0	0	0
23	4	3	5	13	0	0	0	0
24	12	6	3	5	0	0	0	0
24.5	10	4	2	4	0	0	0	0
25	9	3	5	4	97	51	39	30

Table 21

TIME TO DEATH AMONG FEMALE RATS EXPOSED TO EO VAPOR IN THE BRRC STUDY

Exposure months	Number of deaths (dose level/ppm)				Number of rats sacrificed at necropsy interval (dose level/ppm)			
	0	10	33	100	0	10	33	100
1	1	0	0	0	0	0	0	0
2	0	0	0	0	0	0	0	0
3	0	0	0	0	1	1	1	1
4	0	0	1	0	0	0	0	0
5	0	0	0	0	0	0	0	0
6	0	0	0	0	0	0	0	0
7	0	0	0	0	20	10	10	10
8	0	0	0	0	0	0	0	0
9	0	0	0	0	0	0	0	0
10	0	0	0	2	0	0	0	0
11	0	0	0	0	0	0	0	0
12	0	0	0	0	20	10	10	10
13	0	0	1	1	0	0	0	0
14	0	0	0	1	3	0	0	0
15	5	2	4	12	0	0	0	0
16	1	1	0	2	0	0	0	0
17	0	2	1	3	0	0	0	0
18	3	1	3	1	40	20	20	20
19	3	4	4	2	0	0	0	0
20	6	0	1	4	0	0	0	0
21	3	1	4	3	0	0	0	0
22	2	1	2	5	0	0	0	0
23	10	8	6	6	0	0	0	0
24	5	3	2	9	21	12	8	8
24.5	1	2	2	2	95	42	40	18

Table 22

ESTIMATED MEAN NUMBER OF MONTHS A RAT IS WITHOUT A SPECIFIED RESPONSE DURING A 25 MONTH OBSERVATION PERIOD — THE ESTIMATED MEAN FREE PERIOD

Dose (ppm)	Mean free period (specified response)		
	Mononuclear cell leukemia in male rats	Peritoneal mesothelioma in male rats	Mononuclear cell leukemia in female rats
1.0	24.25	24.92	24.53
2.5	24.24	24.91	24.51
5.0	24.23	24.89	24.47
10.0	24.20	24.86	24.40
15.0	24.18	24.83	24.33
20.0	24.15	24.79	24.27
25.0	24.12	24.76	24.20
30.0	24.10	24.72	24.14
33.0	24.09	24.70	24.10
40.0	24.05	24.64	24.01
50.0	24.00	24.55	23.89
75.0	23.88	24.30	23.60
100.0	23.77	24.01	23.33

Table 23
ESTIMATED MEAN NUMBER OF
MONTHS A RAT IS ALIVE
DURING A 25 MONTH
OBSERVATION PERIOD — THE
ESTIMATED MEAN FREE
PERIOD

Dose (ppm)	Mean free period (months)	
	Male rats	Female rats
1.0	23.14	23.61
2.5	23.12	23.57
5.0	23.09	23.52
10.0	23.01	23.41
15.0	22.94	23.30
20.0	22.87	23.19
25.0	22.80	23.07
30.0	22.73	22.96
33.0	22.69	22.89
40.0	22.59	22.72
50.0	22.46	22.47
75.0	22.13	21.81
100.0	21.81	21.08

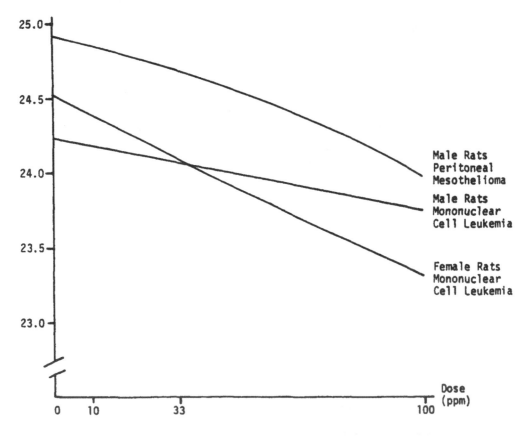

FIGURE 26.   The estimated mean free period (mean number of months without the specified response) based on time to response modeling of the rat data in the BRRC EO inhalation study.

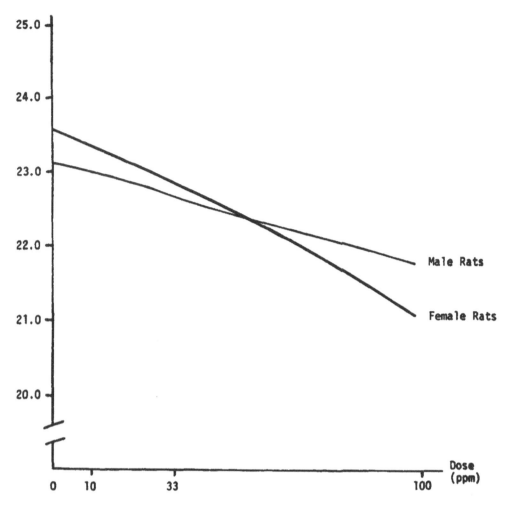

FIGURE 27. The estimated mean number of months without dying (mean free period) based on time to response modeling of the rat data in the BRRC EO inhalation study.

Table 24
DECREASES IN THE ESTIMATED MEAN FREE PERIOD (AVERAGE NUMBER OF MONTHS WITHOUT DYING) BASED ON THE TIME TO RESPONSE MODELING OF THE DATA IN THE BRRC EO STUDY

Dose (ppm)	Decrease in female rat's mean free period	Percentage decrease relative to 25 month lifespan	Corresponding decrease in a 70 year lifespan
0.0	0.0	0.0	0.0
1.0	0.02 months	0.086	0.73 months
5.0	0.11 months	0.44	3.70 months
10.0	0.22 months	0.88	7.39 months
1.54	1.0 day	0.13	1.10 months
0.064	1.0 hr	0.0055	0.05 months
1.38	0.03 months	0.12	1.0 month
0.32	0.007 months	0.027	1.0 week
0.046	0.001 months	0.0039	1.0 day
0.0019	0.00004 months	0.00016	1.0 hr

REFERENCES

1. International Agency for Research on Cancer, Long-term and short-term screening assays for carcinogens: a critical appraisal, in *Monographs on the Evaluation of the Carcinogenic Risk of Chemicals to Humans,* Suppl. 2, Lyon, France, 1980, 318.

2. Occupational Exposure to Ethylene Oxide, 29 CFR Part 1910, Docket No. H-200. Notice of Proposed Rule, Federal Register 48(78): 17284-17319, Occupational Safety and Health Administration, Department of Labor, April 21, 1983.

3. Osterman-Golkar, S., Farmer, P. B., Segerbäck, D., Bailey, E., Calleman, C. J., Svensson, K., and Ehrenberg, L., Dosimetry of ethylene oxide in the rat by quantitation of alkylated histidine in hemoglobin, *Teratogen., Carcinogen. Mutagen.,* 3, 395, 1983.

4. Ehrenberg, L., Hiesche, K. D., Osterman-Golkar, S., and Wennberg, I., Evaluation of genetic risks of alkylating agents: tissue doses in the mouse from air contaminated with ethylene oxide, *Mutat. Res.,* 45, 175, 1974.

5. Osterman-Golkar, S., Ehrenberg, L., Segerbäck, D., and Hällström, I., Evaluation of genetic risks of alkylating agents. II. Hemoglobin as a dose monitor, *Mutat. Res.,* 34, 2, 1976.

6. Calleman, C. J., Ehrenberg, L., Jansson, B., Osterman-Golkar, S., Segerbäck, D., Svensson, K., and Wachtmeister, C. A., Monitoring and risk assessment by means of alkyl groups in hemoglobin in persons occupationally exposed to ethylene oxide, *J. Environ. Pathol. Toxicol.,* 2, 427, 1978.

7. Lawley, P. D. and Jarman, M., Alkylation by propylene oxide of deoxyribonucleic acid, adenine, guanosine and deoxyguanylic acid, *Biochem. J.,* 126, 893, 1972.

8. Osterman-Golkar, S., Ehrenberg, K., and Wachtmeister, C. A., Reaction kinetics and biological action in barley of mono-functional methanesulfonic esters, *Radiat. Bot.,* 10, 303, 1970.

9. Loveless, A., Possible relevance of 0-6 alkylation of deoxyguanosine to mutagenicity of nitrosamines and nitrosamides, *Nature (London),* 223, 206, 1969.

10. Spears, C. P., Nucleophilic selectivity ratios of model and clinical alkylating agents by 4-(4'-nitrobenzyl) pyridine competition, *Mol. Pharmacol.,* 19, 496, 1981.

11. Turchi, G., Bonatti, S., Citti, L., Gervasi, P. G., and Abbondandolo, A., Alkylating properties and genetic activity of 4-vinylcyclo-hexene metabolites and structurally related epoxides, *Mutat. Res.,* 83, 419, 1981.

12. Vogel, E. and Natarajan, A. T., The relation between reaction kinetics and mutagenic action of monofunctional alkylating agents in higher eucaryotic systems. I. Recessive lethal mutations and translocations in Drosophila, *Mutat. Res.,* 62, 51, 1979a.

13. Vogel, E. and Natarajan, A. T., The relation between reaction kinetics and mutagenic action of monofunctional alkylating agents in higher eucaryotic systems. II. Total and partial sex-chromosome loss in Drosophila, *Mutat. Res.,* 62, 101, 1979b.

14. Farmer, P. B., Bailey, E., Lamb, J. H., and Connors, T. A., Approach to the quantitation of alkylated amino acids in haemoglobin by gas chromatography mass spectrometry, *Biomed. Mass Spectrom.,* 7, 41, 1980.

15. Farmer, P. B., Gorf, S. M., and Bailey, E., Determination of hydroxypropylhistidine in haemoglobin as a measure of exposure to propylene oxide using high resolution gas chromatography mass spectrometry, *Biomed. Mass Spectrom.,* 9, 69, 1982.

16. Truong, L., Ward, J. B., Jr., and Legator, M. S., Detection of alkylating agents by the analysis of amino acid residues in hemoglobin and urine. I. The in vivo and in vitro effects of ethyl methanesulfonate, methyl methanesulfonate, hycanthone methanesulfonate, and naltrexone, *Mutat. Res.,* 54, 271, 1978.

17. Pereira, M. A. and Chang, L. W., Binding of chloroform to mouse and rat hemoglobin, *Chem. Biol. Interact.,* 39, 89, 1982.

17a. Green, L. C., Skipper, P. L., Turesky, R. J., Bryant, M. S., and Tannenbaum, S. R., In vivo dosimetry of 4-aminobiphenyl in rats via a cysteine adduct in hemoglobin, *Cancer Res.,* 44, 4254, 1984.

17b. Tannenbaum, S. R. and Skipper, P. L., Biological aspects to the evaluation of risk: dosimetry of carcinogens in man, *Fundam. Appl. Toxicol.,* 4, S367, 1984.

17c. Neumann, H. -G., Analysis of hemoglobin as a dose monitor for alkylating and arylating agents, *Arch. Toxicol.,* 56, 1, 1984.

17d. Shugart, L., Quantitating exposure to chemical carcinogens: in vivo alkylation of hemoglobin by benzo(a)pyrene, *Toxicology,* 34, 211, 1985.

18. Cumming, R. B., Sega, G. A., Horton, C. Y., and Olson, W. H., Degree of alkylation of DNA in various tissues of the mouse following inhalation exposure to ethylene oxide, *Environ. Mutagen.,* 3, 343, 1981.

19. Sega, G. A., Cumming, R. B., Owens, J. G., Horton, C. Y., and Lewis, L. A., Alkylation pattern in developing mouse sperm, sperm DNA and protamine after inhalation of ethylene oxide, *Environ. Mutagen.,* 3, 371, 1981.

20. Sega, G. A., Personal communication, 1983.
20a. Sega, G. A., DNA repair in spermatocytes and spermatids of the mouse, in *Banbury Report 13 Indicators of Genotoxic Exposure*, Bridges, B. A., Butterworth, B. E., and Weinstein, I. B., Eds., Cold Spring Harbor Laboratory, Cold Spring Harbor, N.Y., 1982, 503.
21. Ehrenberg, L., Osterman-Golkar, S., Segerbäck, D., Svensson, K., and Calleman, C. J., Evaluation of genetic risks of alkylating agents. III. Alkylation of haemoglobin after metabolic conversion of ethene to ethene oxide in vivo, *Mutat. Res.*, 45, 175, 1977.
21a. Osterman-Golkar, S., Tissue doses in man: implications in risk assessment, in *Developments in the Science and Practice of Toxicology*, Hayes, A. W., Schnell, R. C., and Miya, T. S., Eds., Elsevier, Amsterdam, 1983, 289.
22. Ehrenberg, L. and Holmberg, B., Extrapolation of carcinogenic risk from animal experiments to man, *Environ. Health Persp.*, 22, 33, 1978.
23. Osterman-Golkar, S., Dosimetry of electrophilic compounds by means of their reaction products with hemoglobin: a method directly applicable to man, in *Health Risk Analysis, Proceedings of the Third Life Sciences Symposium*, Franklin, Gatlinburg, Tenn, 1981, 156.
24. Osterman-Golkar, S. and Ehrenberg, L., Covalent binding of reactive intermediates to hemoglobin as an approach for determining the metabolic activation of chemicals — ethylene, *Drug Metabol. Rev.*, 13, 647, 1982.
25. Djalali-Behzad, G., Hussain, S., Osterman-Golkar, S., and Segerbäck, D., Estimation of genetic risks of alkylating agents. VI. Exposure of mice and bacteria to methyl bromide, *Mutat. Res.*, 84, 1, 1981.
26. Walles, S. A. S., Reaction of benzyl chloride with haemoglobin and DNA in various organs of mice, *Toxicol. Lett.*, 9, 379, 1981.
27. Snellings, W. M., Weil, C. S., and Maronpot, R. R., Final Report on Ethylene Oxide Two-Year Inhalation Study in Rats, Project Report No. 44-20, Bushy Run Research Center, Pa., 1981.
27a. Snellings, W. M., Weil, C. S., and Maronpot, R. R., A two-year inhalation study of the carcinogenic potential of ethylene oxide in Fischer 344 rats, *Toxicol. Appl. Pharmacol.*, 75, 105, 1984.
27b. Snellings, W. M., Weil, C. S., and Maronpot, R. R., A subchronic inhalation study on the toxicologic potential of ethylene oxide in B6C3F$_1$ mice, *Toxicol. Appl. Pharmacol.*, 76, 510, 1984.
28. Cumming, R. B., Personal communication, 1982.
29. Oesch, F., Mammalian epoxide hydrase: inducible enzymes catalyzing the inactivation of carcinogenic and cytotoxic metabolites derived from aromatic and olefinic compounds, *Xenobiotica*, 3, 305, 1973.
30. Oesch, F., Purification and specificity of a microsomal human epoxide hydrase, *Biochem. J.*, 139, 77, 1974.
31. McChesney, E. W., Golberg, L., Parekh, C. K., Russell, J. C., and Min, B. H., Reappraisal of the toxicology of ethylene glycol. II. Metabolism studies in laboratory animals, *Food Cosmet. Toxicol.*, 9, 21, 1971.
32. Gessner, P. K., Parke, D. V., and Williams, R. T., Studies in detoxification. The metabolism of ^{14}C-labelled ethylene glycol, *Biochem. J.*, 79, 482, 1961.
33. Clay, K. L. and Murphy, R. C., On the metabolic acidosis of ethylene glycol intoxication, *Toxicol. Appl. Pharmacol.*, 39, 39, 1977.
34. Jones, A. R. and Wells, G., The comparative metabolism of 2-bromoethanol and ethylene oxide in the rat, *Xenobiotica*, 11, 763, 1981.
35. Blair, A. H. and Vallee, B. L., Some catalytic properties of human liver alcohol dehydrogenase, *Biochemistry*, 5, 2026, 1966.
36. Johnson, M. K., The influence of some aliphatic compounds on rat liver glutathione levels, *Biochem. Pharmacol.*, 14, 1383, 1965.
37. Johnson, M. K., Studies on glutathione S-alkyltransferase of the rat, *Biochem. J.*, 98, 44, 1966.
38. Johnson, M. K., Metabolism of chloroethanol in the rat, *Biochem. Pharmacol.*, 16, 185, 1967a.
39. Johnson, M. K., Detoxification of ethylene chlorohydrin, *Food Cosmet. Toxicol.*, 5, 449, 1967b.
40. Yllner, S., Metabolism of 1,2-dichloroethane-^{14}C in the mouse, *Acta Pharmacol. Toxicol.*, 30, 257, 1971.
41. Grunow, W. and Altmann, H. J., Toxicokinetics of chloroethanol in the rat after single oral administration, *Arch. Toxicol.*, 49, 275, 1982.
42. Green, T. and Hathway, D. E., The chemistry and biogenesis of the S-containing metabolites of vinyl chloride in rats, *Chem. Biol. Interact.*, 17, 137, 1977.
43. Watanabe, P. G., McGowan, G. R., and Gehring, P. J., Fate of [^{14}C] vinyl chloride after single oral administration in rats, *Toxicol. Appl. Pharmacol.*, 36, 339, 1976.
44. Müller, G., Heger, M., and Norpoth, K., Bestimmung der Hydroxyäthylmercaptursäure in Harn Vinylchlorid-Exponierter. Methodische Erfahrungen und analytische Ergebnisse, in *Bericht über die 20. Jahrestagung der deutschen Gesellschaft für Arbeitsmedizin e*, Vol. 5 Gentner-Verlag, Stuttgart, 533, 1980.

45. Rogers, K. M. and Barnsley, E. A., The metabolism of S-carboxyalkyl cysteines in man, *Xenobiotica*, 7, 409, 1977.

46. Waring, R. H., The metabolism of S-carboxymethylcysteine in rodents, marmosets and humans, *Xenobiotica*, 8, 265, 1978.

47. Yllner, S., Metabolism of chloroacetate-1-^{14}C in the mouse, *Acta Pharmacol. Toxicol.*, 30, 69, 1971.

48. Tyler, T. R., Summary Concerning Species Differences in the Metabolism of Ethylene Oxide, memorandum dated September 13, 1979, Carnegie Mellon University, Pittsburgh, Pa., 1979.

49. McKelvey, J. A. and Tyler, T. R., A closed recirculating inhalation metabolism chamber for radiotracer studies, *Toxicol. Appl. Pharmacol.*, 47, A6, 1979.

50. Appelgren, L. E., Eneroth, G., and Grant, C., Studies on ethylene oxide: whole body autoradiography and dominant-lethal test in mice, *Proc. Eur. Soc. Toxicol.*, 18, 315, 1977.

51. Tyler, T. R. and McKelvey, J. A., Dose dependent disposition of ^{14}C labeled ethylene oxide in rats, *Toxicol. Appl. Pharmacol.*, 52, A58, 1980.

52. Tyler, T. R. and McKelvey, J. A., The effect of pre-exposure on the subsequent disposition of ethylene oxide (EO) in rats, *Fed. Proc.*, 39, 749, 1980.

53. Martis, L., Kroes, R., Darby, T. D., and Woods, E. F., Disposition kinetics of ethylene oxide, ethylene glycol, and 2-chloroethanol in the dog, *J. Toxicol. Environ. Health*, 10, 847, 1982.

54. Martis, L., Kroes, R., Brown, R. C., Huang, K. S., and Darby, T. D., Kinetics of ethylene oxide, 2-chloroethanol, and ethylene glycol in the dog, *Fed. Proc.*, 38, 743, 1979.

55. Cawse, J. N., Henry, J. P., Swartzlander, M. W., and Wadia, P. H., Ethylene oxide, in *Kirk-Othmer Encyclopedia of Chemical Technology*, Vol. 9, 3rd ed., Wiley-Interscience, New York, 1980, 432.

56. Thiess, A. M., Observations concerning intoxication due to ethylene oxide exposure, *Arch. Toxicol.*, 20, 127, 1963.

57. Taylor, J. S., Dermatologic hazards from ethylene oxide, *Cutis*, 19, 189, 1977.

58. Sexton, R. J. and Henson, E. V., Experimental ethylene oxide human skin injuries, *Arch. Indust. Hyg. Occup. Med.*, 2, 549, 1950.

58a. Lindhop, C. R., Wilcox, T. W., McKegg, P. M., and Harrie, E. A., Exposure of patients to ethylene oxide during cardiopulmonary bypass using gas-sterilized pump components, *J. Thorac. Cardiovasc. Surg.*, 79, 845, 1980.

58b. Nicholls, A. J. and Platts, M. M., Anaphylactoid reactions due to haemodialysis, haemofiltration, or membrane plasma separation, *Br. Med. J.*, 285, 1607, 1982.

58c. Popli, S., Ing, T. S., Aaugirdas, J. T., Kheirbek, A. O., Geoffrey, W. V., Kilbar, R. M., and Gandhi, V. C., Severe reactions to cuprophan capillary dialyzers, *Artif. Organs*, 6, 312, 1982.

58d. Poothullil, J., Shimizu, A., Day, R. P., and Dolovich, J., Anaphylaxis from the product(s) of ethylene oxide gas, *Ann. Intern. Med.*, 82, 58, 1975.

58e. Dolovich, J. and Bell, B., Allergy to product(s) of ethylene oxide gas, *J. Allergy Clin. Immunol.*, 62, 30, 1978.

58f. Grammer, L. C., Roberts, M., Nicholls, A. J., Platts, M. M., and Patterson, R., IgE against ethylene oxide-altered human serum albumin in patients who have had acute dialysis reactions, *J. Allergy Clin. Immunol.*, 74, 544, 1984.

58g. Marshall, C., Shimizu, A., Smith, E. K. M., and Dolovich, J., Ethylene oxide allergy in a dialysis center: prevalence in hemodialysis and peritoneal dialysis populations, *Clin. Nephrol.*, 21, 346, 1984.

58h. Meynadier, J. M., Guillot, B., Issautier, R., Flavier, J. L., and Meynadier, J., Immediate allergic reaction to EO in a haemodialysis patient, *Presse Med.*, 14, 1245, 1985.

58i. Marshall, C. P., Pearson, F. C., Sagona, M. A., Lee, W., Wathen, R. L., Ward, R. A., and Dolovich, J., Reactions during hemodialysis caused by allergy to ethylene oxide gas sterilization, *J. Allergy Clin. Immunol.*, 75, 563, 1985.

59. Gross, J. A., Haas, M. L., and Swift, T. R., Ethylene oxide neurotoxicity: report of four cases and review of the literature, *Neurology*, 29, 978, 1979.

59a. Jay, W. M., Swift, T. R., and Hull, D. S., Possible relationship of ethylene oxide exposure to cataract formation, *Am. J. Ophthalmol.*, 93, 727, 1982.

59b. Finelli, P. F., Morgan, T. F., Yaar, I., and Granger, C. V., Ethylene oxide-induced polyneuropathy. A clinical and electrophysiologic study, *Arch. Neurol.*, 40, 419, 1983.

59c. Kuzuhara, S., Kanazawa, I., Nakanishi, T., and Egashira, T., Ethylene oxide polyneuropathy, *Neurology*, 33, 377, 1983.

59d. Schröder, J. M., Hoheneck, M., Weis, J., and Deist, H., Ethylene oxide polyneuropathy: clinical follow-up study with morphometric and electron microscopic findings in a sural nerve biopsy, *J. Neurol.*, 232, 83, 1985.

60. Hine, C., Rowe, V. K., White, E. R., Darmer, K. I., Jr., and Youngblood, G. T., Epoxy compounds, in *Patty's Industrial Hygiene and Toxicology*, Vol. 2A, 3rd rev. ed., Clayton, G. D. and Clayton, F. E., Eds., John Wiley & Sons, New York, 1981, 2141.

61. Glaser, Z. R., Ethylene oxide: toxicology review and field study results of hospital use, *J. Environ. Pathol. Toxicol.*, 2, 173, 1979.
62. Jacobson, K. H., Hackley, E. B., and Feinsilver, L., The toxicity of inhaled ethylene oxide and propylene oxide vapors, *AMA Arch. Ind. Health*, 13, 237, 1956.
63. Hollingsworth, R. L., Rowe, V. K., Oyen, F., McCollister, D. D., and Spencer, H. C., Toxicity of ethylene oxide determined on experimental animals, *AMA Arch. Ind. Health*, 13, 217, 1956.
64. Bogyo, D. A., Lande, S. S., Meylan, W. M., Howard, P. H., and Santodonato, J., Investigation of Selected Potential Environmental Contaminants: Epoxides, report No. EPA-560/11-80-005, Office of Toxic Substances, U.S. Environmental Protection Agency, U.S. Department of Commerce, National Technical Information Service, 1980.
65. Proctor, N. H. and Hughes, J. P., *Chemical Hazards of the Workplace*, Lippincott, New York, 1978, 262.
66. Walker, W. J. G. and Greeson, C. E., The toxicity of ethylene oxide, *J. Hyg.*, 32, 409, 1932.
67. Northup, S., Wienckowski, D., Martis, L., and Darby, T., Toxicity caused by acute and subacute intravenous administration of ethylene oxide to the rat, *J. Environ. Pathol. Toxicol.*, 5, 617, 1981.
68. Anger, K., Personal communication, 1982.
69. Sprinz, H., Matzke, H., and Carter, J., Neuropathological Evaluation of Monkeys Exposed to Ethylene and Propylene Oxide, MRI project No. 7222-B, NIOSH contract No. 210-81-6004, report dated February 8, 1982.
70. Snellings, W. M., Ethylene Oxide: Ten- to Eleven-Week Vapor Inhalation Probe Study in Mice, project report No. 45-158, Bushy Run Research Center, Pa., 1982.
71. Irwin, S., Comprehensive observation assessment: in, a systematic quantitative procedure for assessing the behavioural and physiologic state of the mouse, *Psychopharmacologia*, 13, 222, 1966.
72. Sexton, R. J. and Henson, E. V., Dermatologic injuries by ethylene oxide, *J. Ind. Hyg. Toxicol.*, 31, 297, 1949.
73. Garry, V. F., Hozier, F., Jacobs, D., Wade, R. L., and Gray, D. G., Ethylene oxide: evidence of human chromosomal effects, *Environ. Mutagen.*, 1, 375, 1979.
74. Högstedt, C., Rohlen, O., Berndtsson, B. S., Axelson, O., and Ehrenberg, L., A cohort study of mortality and cancer incidence in ethylene oxide production workers, *Br. J. Ind. Med.*, 36, 276, 1979.
75. Morgan, R. W., Claxton, K. W., Divine, B. J., Kaplan, S. D., and Harris, V. B., Mortality among ethylene oxide workers, *J. Occup. Med.*, 23, 767, 1981.
75a. Van Sittert, J. J., de Jong, G., Clare, M. G., Davies, R., Dean, B. J., Wren, L. J., and Wright, A. S., Cytogenetic, immunological and haematological effects in workers in an ethylene oxide manufacturing plant, *Br. J. Ind. Med.*, 42, 19, 1985.
75b. Currier, M. F., Carlo, G. L., Poston, P. L., and Ledford, W. E., A cross sectional study of employees with potential occupational exposure to ethylene oxide, *Br. J. Ind. Med.*, 41, 492, 1984.
75c. Stetka, D. G., Jr., Bleicher, W. T., Jr., and Brewen, J. G., SCE induction is uncoupled from mutation induction in mammalian cells following exposure to ethylnitrosourea (ENU), *Environ. Mutagen.*, 7, 233, 1985.
75d. Soper, K. A., Stolley, P. D., Galloway, S. M., Smith, J. G., Nichols, W. W., and Wolman, S. R., Sister-chromatid exchange (SCE) report on control subjects in a study of occupationally exposed workers, *Mutat. Res.*, 129, 77, 1984.
75e. Stolley, P. D., Soper, K. A., Galloway, S. M., Nichols, W. W., Norman, S. A., and Wolman, S. R., Sister-chromatid exchanges in association with occupational exposure to ethylene oxide, *Mutat. Res.*, 129, 89, 1984.
75f. Gebhart, E., Sister chromatid exchange and structural chromosome aberrations in mutagenicity testing, *Hum. Genet.*, 58, 235, 1981.
75g. Laurent, C., Frederic, J., and Maréchal, F., Augmentation du taux d'échanges entre chromatides-soeurs chez des personnes exposées profesionellement a l'oxyde d'éthylene, *Ann. Genet.*, 26, 138, 1983.
75h. Laurent, C., Frederic, J., and Leonard, A. Y., Sister chromatid exchange frequency in workers exposed to high levels of ethylene oxide, in a hospital sterilization service, *Int. Arch. Occup. Environ. Health*, 54, 33, 1984.
75i. Heflich, R. H., Beranek, D. T., Kodell, R. L., and Morris, S. M., Induction of mutations and sister-chromatid exchanges in Chinese hamster ovary cells by ethylating agents. Relationship to specific DNA adducts, *Mutat. Res.*, 106, 147, 1982.
75j. Morris, S. M., Beranek, D. T., and Heflich, R. H., The relationship between sister-chromatid exchange induction and the formation of specific methylated DNA adducts in Chinese hamster ovary cells, *Mutat. Res.*, 121, 261, 1983.
75k. Natarajan, A. T., Simons, J. W. I. M., Yogel, E. W., and van Zeeland, A. A., Relationship between cell killing, chromosomal aberrations, sister-chromatid exchanges and point mutations induced by monofunctional alkylating agents in Chinese hamster cells. A correlation with different ethylation products in DNA, *Mutat. Res.*, 128, 31, 1984.

75l. Bodell, W. J., Aida, T., and Rasmussen, J., Comparison of sister-chromatid induction caused by nitrosoureas that alkylate or alkylate and crosslink DNA, *Mutat. Res.*, 149, 95, 1985.

75m. Pal, K., The relationship between the levels of DNA-hydrocarbon adducts and the formation of sister-chromatid exchanges in Chinese hamster ovary cells treated with derivatives of polycyclic aromatic hydrocarbons, *Mutat. Res.*, 129, 365, 1984.

75n. Clare, M. G., Dean, B. J., de Jong, G., and van Sittert, N. J., Chromosome analysis of lymphocytes from workers in an ethylene oxide plant, *Mutat. Res.*, 156, 109, 1985.

76. Glaser, Z. R., *Special Occupational Hazard Review with Control Recommendations for the Use of Ethylene Oxide as a Sterilant in Medical Facilities*, DHEW (NIOSH) No. 77-200, U.S. Government Printing Office, Washington, D.C., 1977.

77. Joyner, R. E., Chronic toxicity of ethylene oxide. A study of human responses to long-term low level exposures, *Arch. Environ. Health*, 8, 700, 1964.

78. NIOSH, Toxic and mutagenic effects on ethylene oxide and propylene oxide on the spermatogenic functions in cynomolgus monkeys *(Macaca fascicularis)*, submitted by Environmental Health Research and Testing Inc., Lexington, Ky., NIOSH contract No. 211-81-0024, 1982.

79. Lynch, D. W., Lewis, T. R., Moorman, W. J., Sabharwal, P. S., and Burg, J. A., Toxic and mutagenic effects of ethylene oxide and propylene oxide on spermatogenic functions in Cynomolgus monkeys, *Toxicologist*, 3, 60, 1983.

80. Embree, J. W., Lyon, J. P., and Hine, C. H., The mutagenic potential of ethylene oxide using the dominant-lethal assay in rats, *Toxicol. Appl. Pharmacol.*, 40, 261, 1977.

81. Cumming, R. B. and Michaud, T. A., Mutagenic effects of inhaled ethylene oxide in male mice, *Environ. Mutagen.*, 1, 166, 1979.

82. Generoso, W. M., Cain, K. T., Krishna, M., Sheu, C. W., and Gryder, R. M., Heritable translocation and dominant-lethal mutation induction with ethylene oxide in mice, *Mutat. Res.*, 73, 133, 1980.

82a. Generoso, W. M., Cumming, R. B., Bandy, J. A., and Cain, K. T., Increased dominant-lethal effects due to prolonged exposure of mice to inhaled ethylene oxide, *Mutat. Res.*, 119, 377, 1983.

83. Barlow, S. M. and Sullivan, F. M., An evaluation of animal and human data, in *Reproductive Hazards of Industrial Chemicals*, Academic Press, New York, 1982, 316.

84. Snellings, W. M., Zelenak, J. P., and Weil, C., Effects on reproduction in Fischer 344 rats exposed to ethylene oxide by inhalation for one generation, *Toxicol. Appl. Pharmacol.*, 63, 382, 1982.

85. Yakubova, Z. N., Shamova, H. A., Muftakhova, F. A., and Shilova, L. F., Gynecological disorders in workers engaged in ethylene oxide production, *Kazan. Med. Zh.*, 57, 558, 1976.

86. Hemminki, K., Mutanen, P., Saloniemi, I., Niemi, M. -L., and Vainio, H., Spontaneous abortions in hospital staff engaged in sterilizing instruments with chemical agents, *Br. Med. J.*, 285, 1461, 1982.

86a. Wilcox, A. J. and Horney, L. F., Accuracy of spontaneous abortion recall, *Am. J. Epidemiol.*, 120, 727, 1984.

87. Snellings, W. M., Maronpot, R. R., Zelenak, J. P., and Zaffoon, C. P., Teratology study in Fischer 344 rats exposed to ethylene oxide by inhalation, *Toxicol. Appl. Pharmacol.*, 64, 476, 1982.

88. aBorde, J. B. and Kimmel, C., The teratogenicity of ethylene oxide administered intravenously to mice, *Toxicol. Appl. Pharmacol.*, 56, 16, 1980.

89. Kimmel, C. A., LaBorde, J. B., Jones-Price, C., Ledoux, T. A., and Marks, T. A., Fetal development in New Zealand White (NZW) rabbits treated iv with ethylene oxide during pregnancy, *Toxicologist*, 2 (Abstr.), 249, 1982.

90. Hackett, P. L. et al., Teratogenic Study of Ethylene and Propylene Oxide and n-Butyl Acetate, NIOSH, May 1982, PB83-258038.

91. Ames, B. N., Identifying environmental chemicals causing mutations and cancer, *Science*, 204, 587, 1979.

92. International Agency for Research on Cancer, Ethylene Oxide in IARC Monographs on the Evaluation of the Carcinogenic Risk of Chemicals to Man, Vol. 11, IARC, Lyon, France, 1976, 157.

93. Wolman, S. R., Mutational consequences of exposure to ethylene oxide, *J. Environ. Pathol. Toxicol.*, 2, 1289, 1979.

94. NIOSH, Ethylene oxide (EO): evidence of carcinogenicity, *Curr. Intell. Bull.*, 35, 1981a.

95. Ehrenberg, L. and Hussain, S., Genetic toxicity of some important epoxides, *Mutat. Res.*, 86, 1, 1981.

96. European Chemical Industry Ecology and Toxicology Center (ECETOC), Toxicology of Ethylene Oxide and Its Relevance to Man, technical report No. 5, Brussels, Belgium, 1982.

97. Quint, J., The Toxicity of Ethylene Oxide with Emphasis on Carcinogenic, Reproductive and Genetic Effects, California State Hazard Evaluation System and Information Service, Berkeley, 1982.

98. Embree, J. W., Mutagenicity of Ethylene Oxide and Associated Health Hazard, Ph.D. dissertation, University of California, San Francisco, 1976.

99. Pfeiffer, E. H. and Dunkelberg, H., Mutagenicity of ethylene oxide and propylene oxide and of glycols and halohydrins formed from them during the fumigation of foodstuffs, *Food Cosmet. Toxicol.*, 18, 115, 1980.

100. Tanooka, H., Application of *Bacillus subtilis* spores in the detection of gas mutagens, a case of ethylene oxide, *Mutat. Res.*, 64, 433, 1979.

101. Jones, L. A. and Adams, D. M., Mutations in Bacillus subtilis var. niger (Bacillus globigic) spores induced by ethylene oxide, in *Sporulation Germination, Proc. Int. Spore Conf. 8th, 1980,* Levinson, H. S., Sonenshein, A. L., and Tipper, D. J., Eds., Am. Soc. Microbiol., Washington, D.C., 1981, 273.

102. Cookson, M. J., Sims, P., and Grover, P. L., Mutagenicity of epoxides of polycyclic hydrocarbons correlates with carcinogenicity of parent hydrocarbons, *Nature (London) New Biol.*, 234, 186, 1971.

103. Kölmark, H. G. and Kilbey, B. J., Kinetic studies of mutation induction by epoxides in *Neurospora crassa, Mol. Gen. Genet.*, 101, 89, 1968.

104. Kilbey, B. J. and Kölmark, H. G., A mutagenic after-effect associated with ethylene oxide in *Neurospora crassa, Mol. Gen. Genet.*, 101, 185, 1968.

105. Ehrenberg, L., Gustafsson, A., and Lundqvist, U., Chemically induced mutation and sterility in barley, *Acta Chem. Scand.*, 10, 492, 1956.

106. Ehrenberg, L., Gustafsson, A., and Lundqvist, U., The mutagenic effects of ionizing radiations and reactive ethylene derivatives in barley, *Hereditas*, 45, 351, 1959.

107. Sulovska, K., Lindgren, D., Eriksson, G., and Ehrenberg, L., The mutagenic effect of low concentrations of ethylene oxide in air, *Hereditas*, 62, 264, 1969.

108. MacKey, J., Mutagenesis in *Vulgare* wheat, *Hereditas*, 59, 505, 1968.

109. Jana, M. K. and Roy, K., Effectiveness and efficiency of ethylmethanesulphonate and ethylene oxide for induction of mutations in mice, *Mutat. Res.*, 28, 211, 1975.

110. Smith, H. H. and Lofty, T. A., Comparative effects of certain chemicals on *Tradescantia* chromosomes as observed at pollen tube mitosis, *Am. J. Bot.*, 41, 589, 1954.

111. Bird, M. J., Chemical production of mutations in Drosophila: comparison of techniques, *J. Genet.*, 50, 480, 1952.

112. Nakao, Y. and Auerbach, C., Test of a possible correlation between crosslinking and chromosome breaking abilities of chemical mutagens, *Z. Vererbungs.*, 92, 457, 1961.

113. Fahmy, M. J. and Fahmy, O. G., Cytogenetic analysis of the active carcinogens and tumor inhibitors in *Drosophila melanogaster*. V. Differential genetic response to the alkylating mutagens and X-radiation, *J. Genet.*, 54, 146, 1956.

114. Tan, E. L., Cumming, R. B., and Hsie, A. W., Mutagenicity and cytotoxicity of ethylene oxide in the CHO/HGPRT system, *Environ. Mutagen.*, 3, 683, 1981.

115. Krell, K., Jacobson, E. D., and Selby, K., Mutagenic effect of L5178Y mouse lymphoma cells by growth in ethylene oxide-sterilized polycarbonate flasks, *In Vitro*, 15, 326, 1979.

116. Pero, R. W., Widegren, B., Högstedt, B., and Mitelman, F., *In vivo* and *in vitro* ethylene oxide exposure of human lymphocytes assessed by chemical stimulation of unscheduled DNA synthesis, *Mutat. Res.*, 83, 271, 1981.

117. Pero, R. W., Bryngelsson, T., Widegren, B., Högstedt, B., and Welinder, H., A reduced capacity for unscheduled DNA synthesis in lymphocytes from individuals exposed to propylene oxide and ethylene oxide, *Mutat. Res.*, 104, 193, 1982.

118. Latt, S. A., Allen, J., Bloom, S. E., Carrano, A., Falke, E., Kram, D., Schneider, E., Schreck, R., Tice, R., Whitfield, B., and Wolff, S., Sister chromatid exchanges: a report of the Gene-Tox program, *Mutat. Res.*, 87, 17, 1981.

119. Lamburt, B., Lindblad, A., Holmberg, K., and Francesconi, D., The use of sister chromatid exchange to monitor human populations for exposure to toxicologically harmful agents, in *Sister-Chromatid Exchange*, Wolff, S., Ed., John Wiley & Sons, New York, 1982, 149.

120. Star, E. G., Mutagenic and cytotoxic effect of ethylene oxide on human cell structures, *Zbl. Bakt. Hyg.*, 170, 548, 1980.

121. Yager, J. W. and Benz, R. D., Sister chromatid exchanges induced in rabbit lymphocytes by ethylene oxide after inhalation exposure, *Environ. Mutagen.*, 4, 121, 1982.

121a. Högstedt, B., Gullberg, B., Hedner, K., Kilnig, A. -M., Mitelman, F., Skerfving, S., and Widegren, B., Chromosome aberrations and micronuclei in bone marrow cells and peripheral blood lymphocytes in humans exposed to ethylene oxide, *Hereditas*, 98, 105, 1983.

121b. Sarto, F., Cominato, I., Pinton, A. M., Brovedani, P. G., Faccioli, C. M., Bianchi, V., and Levis, A. G., Cytogenetic damage in workers exposed to ethylene oxide, *Mutat. Res.*, 138, 185, 1984.

121c. Hedner, K., Mitelman, F., and Pero, R. W., Sister-chromatid exchanges in human lymphocytes after a non-S-phase incubation period to allow excision DNA repair — in vitro exposure to N-acetoxy-2-acetylaminofluorene and ethylene oxide, *Mutat. Res.*, 129, 71, 1984.

122. Lynch, D. W., Lewis, T. R., and Moorman, W. J., Chronic inhalation toxicity of ethylene oxide and propylene oxide in rats and monkeys — a preliminary report, *Toxicologist*, 2, 11, 1982.

122a. Lynch, D. W., Lewis, T. R., Moorman, W. J., Burg, T. R., Groth, D. H., Kahn, A., Ackerman, L. J., and Cockrell, B. Y., Carcinogenic and toxicologic effects of inhaled ethylene oxide and propylene oxide in F344 rats, *Toxicol. Appl. Pharmacol.*, 76, 69, 1984.

123. NIOSH, Effect of Ethylene Oxide and Propylene Oxide on the Induction of Chromosomal Aberrations and Sister Chromatid Exchanges in Cynomolgus Monkey *(Macaca fascicularis)* Lymphocytes, submitted by Environmental Health Research and Testing Inc., Lexington, Ky, NIOSH contract No. 211-81-0024, 1981b.

124. Jenssen, D. and Ramel, C., The micronucleus test as part of a short-term mutagenicity test program for the prediction of carcinogenicity evaluated by 143 agents tested, *Mutat. Res.*, 75, 191, 1980.

125. Appelgren, L. E., Eneroth, G., Grant, C., Landstrom, L. E., and Tenghagen, K., Testing of ethylene oxide for mutagenicity using the micronucleus test in mice and rats, *Acta Pharmacol. Toxicol.*, 43, 69, 1978.

126. Radman, M., Jeggo, P., and Wagner, R., Chromosomal rearrangements and carcinogenesis, *Mutat. Res.*, 98, 249, 1982.

127. Strekalova, Z. E., On the problem of the mutagenic effect of ethylene oxide on mammals, *Toksikol. Nov. Prom. Khim. Veshchestv.*, 12, 72, 1971.

128. Strekalova, Z. E., Chirkova, E. M., and Golubovich, E., Mutagenic action of ethylene oxide on sex and somatic cells in male white rats, *Toksikol. Nov. Prom. Khim. Veshchestv.*, 14, 11, 1975.

129. Poirier, V. and Papadopoulo, D., Chromosomal aberrations induced by ethylene oxide in a human amniotic cell line *in vitro*, *Mutat. Res.*, 104, 255, 1982.

130. Ehrenberg, L., Risk assessment of ethylene oxide and other compounds, in *Assessing Chemical Mutagens: The Risk to Humans, Banbury Report 1*, Cold Spring Harbor Laboratory, Cold Spring Harbor, New York, 1979, 157.

131. Reyniers, J. A., Sacksteder, M. R., and Ashburn, L. L., Multiple tumors in female germ-free inbred albino mice exposed to bedding treated with ethylene oxide, *J. Natl. Cancer Inst.*, 32, 1045, 1964.

132. Allen, R. C., Meier, H., and Hoag, W. G., Ethylene glycol produced by ethylene oxide sterilization and its effects on blood-clotting factors in an inbred strain of mice, *Nature (London)*, 193, 387, 1962.

133. Lawley, P. D., Carcinogenesis by alkylating agents, in *Chemical Carcinogens*, Searle, C. E., Ed., American Chemical Society, Washington, D.C., 1976, 176.

134. Fishbein, L., Alkylating agents — epoxides and lactones, in *Studies in Environmental Science*, Vol. 4, Elsevier, Amsterdam, 1979, 93.

135. Department of Labor, Occupational Safety and Health Administration, Advance notice of proposed rulemaking, *Fed. Regis.*, 47(17), 3566, 1982.

136. Walker, A. I. T., Thorpe, E., and Stevenson, D. E., The toxicology of dieldrin (HEOD). I. Long-term oral toxicity studies in mice, *Food Cosmet. Toxicol.*, 11, 415, 1972.

137. Bär, F. and Griepentrog, F., Langzeitfutterungsversuch an Ratten mit Athylenoxidbegastem Futter, *Bundesgesundheitsblatt*, 12, 106, 1969.

138. Manchon, P. H., Buquet, A., and Atteba, S., Toxicologie chronique du pain conservé par l'oxyde d'éthylène, *Food Cosmet. Toxicol.*, 8, 17, 1970.

139. Thomas, J. A., Lambre, C. and Henry, M., Accroissement et modification du pouvoir carcinogène du virus des tumeurs mammaires, après traitement par l'oxyde d'éthylène; repartition des particles virales, *C.R. Acad. Sci. Paris*, 273, 244, 1971.

140. Walpole, A. L., Carcinogenic action of alkylating agents, *Ann. N.Y. Acad. Sci.*, 68, 750, 1958.

141. Van Duuren, B. L., Orris, L., and Nelson, N., Carcinogenicity of epoxides, lactones, and peroxy compounds. II, *J. Natl. Cancer Inst.*, 35, 707, 1965.

142. Dunkelberg, H., On the oncogenic activity of ethylene oxide and propylene oxide in mice, *Br. J. Cancer*, 39, 588, 1979.

143. Dunkelberg, H., Kanzerogene Aktivitat von Ethylenoxid und seinen Reaktions-produkten 2-Chloroethanol, 2-Bromoethanol, Ethylenglykol und Diethylenglykol. I. Kanzerogenität von Ethylenoxid im Vergleich zu 1,2-Propylenoxid bei Subkutaner Applikation an Maüsen, *Zbl. Bakt. Hyg.*, 174, 383, 1981.

144. Dunkelberg, H., Carcinogenicity of ethylene oxide and 1,2-propylene oxide upon intragastric administration to rats, *Br. J. Cancer*, 46, 924, 1982.

145. Food and Drug Administration, Nonclinical laboratory studies — good laboratory practice regulations, *Fed. Regis.*, 43(247), 59986, 1978.

146. Union Carbide Corporation, Supplemental Report to the Environmental Protection Agency, April 4, 1983.

146a. Garman, R. H., Snellings, W. M., and Maronpot, R. R., Brain tumors in F344 rats associated with chronic inhalation exposure to ethylene oxide, *Neurotoxicol.*, 6, 117, 1985.

147. Russell, L. B., Cumming, R. B., and Hunsicker, P. R., Specific-locus mutation rates in the mouse following inhalation of ethylene oxide, and application of the results to estimation of human genetic risk, *Mutat. Res.*, 129, 381, 1984.

148. Ad Hoc Panel on Chemical Carcinogenesis Testing and Evaluation, final report, Aug. 17, National Toxicology Program, 1984, 276.

149. Ito, N., Hagiwara, A., Shibata, M., Ogiso, T., and Fukushima, S., Induction of squamous cell carcinoma in the forestomach of F344 rats treated with butylated hydroxyanisole, *Gann*, 73, 332, 1982.

149a. Ito, N., Fukushima, S., Hagiwara, A., Shibata, M., and Ogiso, T., Carcinogenicity of butylated hydroxyanisole in F344 rats, *J. Natl. Cancer Inst.*, 70, 343, 1983.

149b. Nera, E. A., Lok, E., Iverson, F., Ormsby, E., Karpinski, K. F., and Clayson, D. B., Short-term pathological and proliferative effects of butylated hydroxyanisole and other phenolic antioxidants in the forestomach of Fischer 344 rats, *Toxicology*, 32, 197, 1984.

150. Orme, T., Carcinogenesis bioassay of ethylene oxide, *Toxicol. Res. Proj. Directory*, 6(4), 1981.

151. Andersen, S. R., Ethylene oxide toxicity. A study of tissue reactions to retained ethylene oxide, *J. Lab. Clin. Med.*, 77, 346, 1971.

152. Grasso, P. and Golberg, L., Subcutaneous sarcoma as an index of carcinogenic activity, *Food Cosmet. Toxicol.*, 4, 297, 1966.

153. Grasso, P., Gangolli, S. D., and Hooson, J., Connective tissue response to a short-term series of subcutaneous injections of sorbic acid or aflatoxin. Physico-chemical factors determining reaction to sorbic acid, *Br. J. Cancer*, 23, 787, 1969.

154. Grasso, P., Gangolli, S. D., Golberg, L., and Hooson, J., Physiochemical and other factors determining local sarcoma production by food additives, *Food Cosmet. Toxicol.*, 9, 463, 1971.

155. Gangolli, S. D., Grasso, P., and Golberg, L., Physical factors determining the early local tissue reactions produced by food colourings and other compounds injected subcutaneously, *Food Cosmet. Toxicol.*, 5, 601, 1967.

156. Gangolli, S. D., Grasso, P., Golberg, L., and Hooson, J., Protein binding by food colorings in relation to the production of subcutaneous sarcoma, *Food Cosmet. Toxicol.*, 10, 449, 1972.

157. Hooson, J., Grasso, P., and Gangolli, S. D., Early reactions of the subcutaneous tissue to repeated injections of carcinogens in aqueous solutions, *Br. J. Cancer*, 25, 505, 1971.

157a. Theiss, J. C., Utility of injection site tumorigenicity in assessing the carcinogenic risk of chemicals to man, *Regul. Toxicol. Pharmacol.*, 2, 213, 1982.

158. Ward, J. M., Testimony at Hearing on Occupational Exposure to Ethylene Oxide 29 C.F.R. Part 1910, Proposed Rule, 1983, 1106.

159. Dunkelberg, H., Kanzerogene aktivität von ethylenoxid und seinen reaktionsprodukten 2-chlorethanol, 2-bromethanol, ethylenglycol und diethylenglycol, *Zbl. Bakt. Hyg.*, 177, 269, 1983.

160. Ad Hoc Working Group on Oil/Gavage in Toxicology, The Nutrition Foundation, Washington, D.C., 1983.

161. Second Annual Report on Carcinogens, National Toxicology Program, 1981.

162. Weinbren, D., Salm, R., and Greenberg, G., Intramuscular injections of iron coumpounds and oncogenesis in man, *Br. Med. J.*, i, 683, 1978.

163. Tomatis, L., The value of long-term testing for the implementation of primary prevention, in *Origins of Human Cancer*, Book C., Hiatt, H. H., Watson, J. D., and Winsten, J. A., Eds. Cold Spring Harbor Laboratory, Cold Spring Harbor, N.Y., 1977a, 1339.

164. Pelfrene, A. F., Comment on methodology and interpretation of results, *J. Natl. Cancer Inst.*, 58(3), 475, 1977.

165. Tomatis, L., Comment on methodology and interpretation of results, *J. Natl. Cancer Inst.*, 59(5), 1341, 1977b.

166. Alarie, Y., Sensory irritation from airborne chemicals in animals; a basis to establish acceptable levels of exposure, in *Toxicology of the Nasal Passages*, Barrow, C. S., Ed., Hemispere Pulb. Corp., Washington, D.C., 1986, 91.

167. Roe, F. J. C., Testing for carcinogenicity and the problem of pseudocarcinogenicity, *Nature (London)*, 303(5919), 657, 1983.

168. Goodman, D. G., Ward, J. M., Squire, R. A., Chu, K. C., and Linhart, M. S., Neoplastic and nonneoplastic lesions in aging F344 rats, *Toxicol. Appl. Pharmacol.*, 48, 237, 1979.

169. Richmond, G. W., Abrahams, R. H., Nemenzo, J. H., and Hine, C. H., An evaluation of possible effects on health following exposure to ethylene oxide, in *Occupational Health and the Chemical Industry*, Proc. 11th Congress, Orford, R. R., Cowell, J. W., Jamieson, G. G., and Love, E. J., Eds., MEDICHEM Calgary '83 Association, Calgary, Canada, 1984, 47.

169a. Richmond, G. W., Abrahams, R. H., Nemenzo, J. H., and Hine, C. H., An evaluation of possible effects on health following exposure to ethylene oxide, *Arch. Environ. Health*, 40, 20, 1985.

170. Perkins, J. J., *Principles and Methods of Sterilization in Health Sciences*, 2nd ed., Charles C Thomas, Springfield, Illinois, 1969, 502.

171. Kereluk, K. and Lloyd, R. S., *Ethylene Oxide Sterilization, A Current Review of Principles and Practices*, American Sterilizer Company, Erie, Pa., 1974.

172. Bolton, N. and Ketcham, N., Determination of ethylene oxide in air, *AMA Arch. Environ. Health*, 8, 711, 1964.

173. Goldgraben, R. and Zank, N., *Mitigation of Worker Exposure to Ethylene Oxide,* Mitre Corporation Report, McLean, Va., 1981.

174. Qazi, A. H. and Ketcham, N. H., A new method for monitoring personal exposure to ethylene oxide in the occupational environment, *Am. Ind. Hyg. Assoc. J.*, 38, 635, 1977.

175. Pilny, R. J. and Coyne, L. B., Industrial Hygiene Laboratory, Dow Chemical Company, Midland, Mich., 1982, Unpublished data.

176. Potter, W. D., OSHA Analytical Method No. 30, OSHA Analytical Laboratory, Salt Lake City, Utah, 1981.

177. Romano, S. J. and Renner, J. A., Analysis of ethylene oxide-worker exposure, *Am. Ind. Hyg. Assoc. J.*, 40, 8, 1979.

178. Opp, C. W., Monitoring ethylene oxide with direct reading instrumentation, in The Safe Use of Ethylene Oxide, Health Industry Manufacturers Association Report 80-4, Washington, D.C., 1980, 85.

179. Roy, P. A., Engineering control of ethylene oxide exposures arm gas sterilization, in The Safe Use of Ethylene Oxide, Health Industry Manufacturers Association Report 80-4, Washington, D.C., 1980, 196.

179a. Department of Health and Human Services, Centers for Disease Control, National Institute of Occupational Safety and Health; NIOSH/MSHA Testing and Certification of Air Purifying Respirators and End-of-Service-Life Indicators, Notice of Acceptance of Application for Approval of Air Purifying Respirators with End-of-Service-Life Indicators (ESLI), *Fed. Regis.*, 49 (140), 29270, 1984.

180. Kaye, S., Testimony to be presented at informal rulemaking hearing relating to occupational exposure to Ethylene Oxide, 1983.

181. Högstedt, C., Malmqvist, N., and Wadman, B., Leukemia in workers exposed to ethylene oxide, *J. Am. Med. Assoc.*, 241, 1132, 1979a.

182. Thiess, A. M., Frentzel-Beyme, R., Link, R., and Stocker, W. G., Mortality study on employees exposed to alkylene oxides (ethylene oxide/propylene oxide) and their derivatives, Proc. Int. Symp. Prevention Occup. Cancer, Helsinki, Finland, 1981.

183. Ehrenberg, L. and Hällström, T., Hematologic Studies on Persons Occupationally Exposed to Ethylene Oxide, International Atomic Energy Agency Report, SM 92/26, 1967, 326.

184. Abrahams, R. H., Recent studies with workers exposed to ethylene oxide, in The Safe Use of Ethylene Oxide, HIMA Report No. 80-4, 1980, 27.

185. Maugh, T. H., II., Biological markers for chemical exposure, *Science,* 215, 643, 1982.

185a. Richardson, C. R., Howard, C. A., Sheldon, T., Wildgoose, J., and Thomas, M. G., The human lymphocyte in vitro cytogenetic assay: positive and negative control observations on 30000 cells, *Mutat. Res.*, 141, 59, 1984.

186. Johnson and Johnson, Preliminary Report of Pilot Research Chromosome Study of Workers at Sites Where Ethylene Oxide Gas is Used as a Sterilant, 1982.

187. Thiess, A. M., Schwegler, H., Fleig, I., and Stocker, W. G., Mutagenicity study of workers exposed to alkylene oxides (ethylene oxide/propylene oxide) and derivatives, *J. Occup. Med.*, 23, 343, 1981.

188. Wolman, S. R., Some views on cytogenetic effects of ethylene oxide, in The Safe Use of Ethylene Oxide, HIMA Report No. 80-4, 1980, 39.

189. Hansen, J. P., Allen, J., Brock, K., Falconer, J., Helms, M. J., Shaver, G. C., and Strohm, B., Normal sister chromatid exchange levels in hospital sterilization employees exposed to ethylene oxide, *J. Occup. Med.*, 26, 19, 1984.

190. Latt, S. A. and Schreck, R. R., Sister chromatid exchange analysis, *Am. J. Hum. Genet.*, 32, 297, 1980.

191. Hanawalt, P. C., Perspectives on DNA repair and inducible recovery phenomena, *Biochimie*, 61, 847, 1982.

192. Yager, J. W., Hines, C. J., and Spear, R. C., Exposure to ethylene oxide at work increases sister chromatid exchanges in human peripheral lymphocytes, *Science,* 219, 1221, 1983.

193. Preston, J., Occupational Exposure to Ethylene Oxide, 29 C.F.R. part 1910, proposed rule, docket No. H-200, supplemental submission, 1983.

194. Sandberg, A. A., Ed., *Sister Chromatid Exchange,* Liss, New York, 1982.

195. Müller, D., Natarajan, A. T., Obe, G., and Röhrborn, G., Eds., *Sister Chromatid-Exchange Test,* Thieme, Stuttgart, 1982.

196. Wolff, S., Ed., *Sister Chromatid Exchange,* John Wiley & Sons, New York, 1982a.

197. Oikawa, A., Sakai, S., Horaguchi, K., and Tohda, H., Sensitivies of peripheral lymphocytes from healthy humans to induction of sister chromatid exchanges by chemicals, *Cancer Res.*, 43, 439, 1983.
198. Mazrimas, J. A. and Stetka, D. G., Direct evidence for the role of incorporated BUdR in the induction of sister chromatid exchanges, *Exp. Cell Res.*, 117, 23, 1978.
199. Carrano, A. V., Minkler, J. L., Stetka, D. G., and Moore, D. H., Variation in the baseline sister chromatid exchange frequency in human lymphocytes, *Environ. Mutagen.*, 2, 325, 1980.
200. Wolff, S. and Fijtman, N., X-ray sensitization of chromatids with unifilarly and bifilarly substituted DNA, *Mutat. Res.*, 80, 133, 1981.
201. Speit, G., Effects of temperature on sister chromatid exchanges, *Hum. Genet.*, 55, 333, 1980.
202. Gutierrez, C., Schvartzman, J. B., and Lopez-Saez, J. F., Effect of growth temperature on the formation of sister-chromatid exchanges in BrdUrd-substituted chromosomes, *Exp. Cell. Res.*, 134, 73, 1981.
203. Kato, H. and Sandberg, A. A., The effect of sera on sister chromatid exchanges *in vitro*, *Exp. Cell Res.*, 109, 445, 1977.
204. Bianchi, N. O., Bianchi, M. S., and Larramedy, M., Kinetics of human lymphocyte division and chromosomal radiosensitivity, *Mutat. Res.*, 63, 317, 1979.
205. Morgan, W. F. and Crossen, P. E., Factors influencing sister-chromatid exchange rate in cultured human lymphocytes, *Mutat. Res.*, 81, 395, 1981.
206. Wolff, S., Problems and prospects in utilization of cytogenetics to estimate exposure at toxic chemical waste dumps, *Environ. Health Perspect.*, 48, 25, 1983.
207. Wolff, S., Difficulties in assessing the human health effects of mutagenic carcinogens by cytogenetic analyses, *Cytogenet. Cell Genet.*, 33, 7, 1982b.
208. Newbold, R. F., Warren, W., Medcalf, A. S. C., and Amos, J., Mutagenicity of carcinogenic methylating agents is associated with a specific DNA modification, *Nature (London)*, 283, 596, 1980.
209. Heflich, R. H., Beranek, D. T., Kodell, R. L., and Morris, S. M., Induction of mutations and sister-chromatid exchanges in Chinese hamster ovary cells by ethylating agents. Relationship to specific DNA adducts, *Mutat. Res.*, 106, 147, 1982.
210. Goth-Goldstein, R., Repair of DNA damaged by alkylating carcinogens is defective in xeroderma pigmentosum derived fibroblasts, *Nature (London)*, 267, 81, 1977.
211. Day, R. S., Ziolkowski, C. H. J., Scudiero, D. A., Meyer, S. A., Lubiniecki, A. S., Girardi, A. J., Galloway, S. M., and Bynum, G. M., Defective repair of alkylated DNA by human tumour and SV40-transformed human cell strains, *Nature (London)*, 288, 724, 1980.
212. Wolff, S., Relation between DNA repair, chromosome aberrations and sister chromatid exchanges, in *DNA Repair Mechanisms*, Hanawalt, P. C., Friedberg, E. C., and Fox, C. F., Eds., Academic Press, New York, 1978a, 751.
213. Carrano, A. V., Thompson, L. H., Stetka, D. G., Minkler, J. L., Marzimas, J. A., and Fong, S., DNA crosslinking, sister chromatid exchange and specific locus mutations, *Mutat. Res.*, 63, 175, 1979.
214. Duncan, A. M. V. and Evans, H. J., Molecular lesions involved in the induction of sister-chromatid exchange, *Mutat. Res.*, 105, 423, 1982.
215. Painter, R. B., A replication model for sister chromatid exchange, *Mutat. Res.*, 70, 337, 1980.
216. Ishii, Y. and Bender, M. A., Effects of inhibitors of DNA synthesis on spontaneous and ultraviolet light induced sister-chromatid exchanges in Chinese hamster cells, *Mutat. Res.*, 79, 19, 1980.
217. Carrano, A. V. and Thompson, L. H., Sister chromatid exchange and gene mutation, *Cytogenet. Cell Genet.*, 33, 57, 1982.
218. Morris, S. M., Heflich, R. H., Beranek, D. T., and Kodell, R. L., Alkylation-induced sister-chromatid exchanges correlate with reduced cell survival, not mutations, *Mutat. Res.*, 105, 163, 1982.
219. Thilagar, A. and Kumaroo, V., Induction of chromosome damage by methylene chloride in CHO cells, *Mutat. Res.*, 116, 361, 1983.
220. Livingston, G. K. and Fineman, R. M., Correlation of human lymphocyte SCE frequency with smoking history, *Mutat. Res.*, 119, 59, 1983.
220a. Husgafvel-Pursiainen, K., Sorsa, M., Jarventaus, H., and Norppa, H., Sister-chromatid exchanges in lymphocytes of smokers in an experimental study, *Mutat. Res.*, 138, 197, 1984.
221. Armitage, P. and Doll, R., The age distribution of cancer and a multistage theory of carcinogenesis, *Br. J. Cancer*, 8, 1, 1954.
222. Sielken, R., Testimony at Hearing on Occupational Exposure to Ethylene Oxide 29 C.F.R. Part 1910, proposed rule, 1983a, 561.
223. Bridges, B. A., Some general principles of mutagenicity screening and a possible framework for testing procedures, *Environ. Health Perspect.*, 6, 221, 1973.
224. Bridges, B. A., The three-tier approach to mutagenicity screening and the concept of radiation-equivalent dose, *Mutat. Res.*, 26, 335, 1974.

225. Ehrenberg, L., Risk assessment of ethylene oxide and other compounds, in *Banbury Report 1. Assessing Chemical Mutagens: The Risk to Humans*, McElheny, V. K. and Abrahamson, S., Eds., Cold Spring Harbor Laboratory, Cold Spring Harbor, N.Y., 1979a, 157.

226. Segerbäck, D., Calleman, C. J., Ehrenberg, L., Löfroth, G., and Osterman-Golkar, S., Evaluation of genetic risks of alkylating agents. IV. Quantitative determination of alkylated amino acids in haemoglobin as a measure of the dose after treatment of mice with methyl methanesulfonate, *Mutat. Res.*, 49, 71, 1978.

227. Committee 17, Environmental Mutagen Society, Environmental mutagenic hazards, *Science*, 187, 503, 1975.

228. Mengs, V., Lang, W., and Poch, J. A., The carcinogenic action of aristolochic acid in rats, *Arch. Toxicol.*, 51, 107, 1982.

229. Mengs, V., On the histopathogenesis of rat forestomach carcinoma caused by aristolochic acid, *Arch. Toxicol.*, 52, 209, 1983.

230. Koestner, A., Problems with Brain Tumors, Toxicology Forum, Winter Meeting, Washington, D.C., 1983, 294.

231. Jones, A. R. and Edwards, K., Comparative metabolism of ethylene dimethanesulfonate and ethylene dibromide, *Experientia*, 24, 1100, 1968.

232. Hathway, D. E., Mechanisms of vinyl chloride carcinogenicity/mutagenicity, *Br. J. Cancer*, 44, 597, 1981.

233. International Agency for Research on Cancer, Some monomers, plastics and synthetic elastomers, and acrolein, in Monographs on the Evaluation of the Carcinogenic Risk of Chemicals to Humans, Vol. 19, International Agency for Research on Cancer, Lyon, France, 1979, 398.

234. Gehring, P. J. and Blau, G. E., Mechanisms of carcinogenesis: dose response, *J. Environ. Pathol. Toxicol.*, 1, 163, 1977.

235. Anderson, M. W., Hoel, D. G., and Kaplan, N. L., A general scheme for the incorporation of pharmacokinetics in low-dose risk estimation for chemical carcinogenesis: example — vinyl chloride, *Toxicol. Appl. Pharmacol.*, 55, 154, 1980.

236. Schumann, A. M., Watanabe, P. G., Reitz, R. H., and Gehring, P. J., The importance of pharmacokinetic and macromolecular events as they relate to mechanisms of tumorigenicity and risk assessment, in *Toxicology of the Liver*, Plaa, G. and Hewitt, R., Eds., Raven Press, New York, 1982, 311.

237. Hoel, D. G., Kaplan, N. L., and Anderson, M. W., Implication of nonlinear kinetics on risk estimation in carcinogenesis, *Science*, 219, 1032, 1983.

238. Ward, J. M. and Reynolds, C. W., Large granular lymphocyte leukemia. A heterogeneous lymphocytic leukemia in F344 rats, *Am. J. Pathol.*, 111, 1, 1983.

238a. Reynolds, C. W., Bere, E. W., Jr. and Ward, J. M., Natural killer activity in the rat. IV. Characterization of transplantable large granular lymphocyte (LGL) leukemias in the F344 rat, *J. Immunol.*, 132, 534, 1984.

238b. Reynolds, E. W., Ward, J. M., Denn, A. C., III, and Bere, E. W., Jr., Characterization of large granular lymphocyte (LGL) tumors in the rat, in *B and T Cell Tumors*, Vitetta, E. S., Ed., Academic Press, Orlando, Fla., 1982, 51.

238c. Reynolds, C. W., Ward, J. M., Denn, A. C., III, and Bere, E. W., Jr., Identification and characterization of large granular lymphocyte (LGL) leukemias in F344 rats, in *NK Cells and Other Natural Effector Cells*, Herberman, R. B., Ed., Academic Press, Orlando, Fla., 1982, 1161.

238d. Reynolds, C. B., Denn, A. C., III, Bere, E. W., Jr., and Ward, J. M., Characterization of large granular lymphocyte (LGL) tumors in the F344 rat, *Immunobiol.*, 163, 409, 1982.

238e. Reynolds, C. W. and Foon, K. A., T$_\gamma$-lymphoproliferative disease and related disorders in humans and experimental animals: a review of the clinical, cellular and functional characteristics, *Blood*, 64, 1146, 1984.

238f. Stromberg, P. C., Animal model of human disease: large granular cell leukemia in F344 rats. Model for human T$_\gamma$-lymphoma, malignant histiocytosis, and T-cell chronic lymphocytic leukemia, *Am. J. Pathol.*, 119, 517, 1985.

239. Maronpot, R. R., Personal communication, 1982.

240. Itoh, K., Tsuchikawa, K., Awataguchi, T., Shiiba, K., and Kumagai, K., A case of chronic lymphocytic leukemia with properties characteristic of natural killer cells, *Blood*, 61, 940, 1983.

241. Stein, P., Peiper, S., Butler, D., Melvin, S., Williams, D., and Stass, S., Granular acute lymphoblastic leukemia, *Am. J. Clin. Pathol.*, 79, 426, 1983.

242. Whitaker, D., Papadimitriou, J. M., and Walters, M. N. I., The mesothelium and its reactions; a review, *Crit. Rev. Toxicol.*, 10, 81, 1982.

243. Borzelleca, J. F., Hogan, G. K., and Koestner, A., A chronic toxicity/carcinogenicity study of FDXC Blue No. 2 in rats, *Food Chem. Toxicol.*, 23, 551, 1985.

244. Ward, J. M. and Rice, J. M., Naturally occurring and chemically induced brain tumors of rats and mice in carcinogenesis bioassays, *Ann. N.Y. Acad. Sci.*, 381, 304, 1982.

245. Boorman, G. A., Letter to Dr. D. Groth (NIOSH) discussing the analysis by a Pathology Working Group of the pathology report on the two-year inhalation study on ethylene oxide in male F344 rats done by NIOSH, April 21, 1983.

246. Koestner, A., Problems with brain tumors, Toxicology Forum, Winter Meeting, Washington, D.C., 1983, 294.

247. Kleihues, P., Mende, C. H. R., and Reucher, W., Tumors of the peripheral and central nervous system induced in BD rats by prenatal application of methyl methanesulfonate, *Eur. J. Cancer*, 8, 641, 1972.

248. Bigner, D. D. and Swenberg, J. A., *Jänisch and Schreiber's Experimental Tumors of the Central Nervous System*, Upjohn, Kalamazoo, Mich., 1977, 83.

249. Koestner, A., Animal model of human disease: N-nitrosourea-induced neurogenic tumors in the rat, *Comp. Pathol. Bull.*, 18, 2, 1978.

250. Swenberg, J. A., Clenderon, N., Denlinger, R., and Gordon, W. A., Sequential development of ethylnitrosourea-induced neurinomas: morphology, biochemistry and transplantability, *J. Natl. Cancer Inst.*, 55, 147, 1975.

251. Vinores, S. A. and Koestner, A., Reduction of ethynitrosourea induced neoplastic proliferation in rat trigeminal nerves by nerve growth factor, *Cancer Res.*, 42, 1038, 1982.

252. Koestner, A., Aspartame and brain tumors: pathology issues, in *Aspartame: Physiology and Biochemistry*, Stegink, L. D. and Filer, L. J., Jr., Eds., Marcel Dekker, New York, 447, 1984.

253. Weschler, W. and Koestner, A., Developmental biology related to oncology, in *Principles of Surgical Oncology*, Raven, R., Ed., Plenum Press, New York, 93, 1977.

254. Waite, C. P., Patty, F. A., and Yant, W. P., Acute response of guinea pigs to some new commercial organic compounds: ethylene oxide, *Public Health Rep.*, 45, 1832, 1930.

255. Amdur, M. O. and Mead, J., The respiratory response of guinea pigs to inhalation of ethylene oxide, *Arch. Ind. Health.*, 14, 553, 1957.

256. Salinas, E., Sasich, L., Hall, D. H., Kennedy, R. M., and Morriss, M., Acute ethylene oxide intoxication, *Drug Intell. Clin. Pharm.*, 15, 384, 1981.

257. Ewart, B., A case of ethylene oxide poisoning, *J. Swed. Phys.*, 34, 1607, 1937.

258. Lundberg, A., A case of T gas (Ethylene Oxide) poisoning, *Norde Med. Tidskrift*, 16, 1862, 1938.

259. Yager, J. L. W., Sister Chromatid Exchanges, Biological Monitor for Workplace Exposure: A Comparative Study of Exposure-Response to Ethylene Oxide, Doctoral dissertation Report NCOHC82-B-2, the University of California, Berkeley, 1982.

260. Kaplan, E. L. and Meier, P., Nonparametric estimation from incomplete observations, *J. Am. Stat. Assoc.*, 53, 457, 1958.

260a. McNeill, J. M. and Wills, E. D., The formation of mutagenic derivatives of benzo[a]pyrene by peroxidising fatty acids, *Chem. Biol. Interactions*, 53, 197, 1985.

261. Ozawa, N., Yajima, A., Soh, K., Takabayashi, T., Sato, S., Sato, A., and Suzuki, M., Sister chromatid exchanges in human lymphocytes after exposure to pulse-wave ultrasound, *Tohoku J. Exp. Med.*, 143, 473, 1984.

262. Stella, M., Trevisan, K., Montaldi, A., Zaccaria, G., Rossi, G., Bianchi, V., and Levis, A. G., Induction of sister-chromatid exchanges in human lymphocytes exposed in vitro and in vivo to therapeutic ultrasound, *Mutat. Res.*, 138, 75, 1984.

263. Tarone, R. E., Tests for trend in life analyses, *Biometrika*, 62, 679, 1975.

264. Armitage, P., Tests for linear trend in proportions and frequencies, *Biometrics*, 11, 375, 1955.

265. Pott, F., Brockhaus, A., and Huth, F., Untersuchungen zur Kanzerogenitat von polyzyklischen aromatischen Köhlenwasserstoffen in Tierexperiment, *Zentralbl. Bakteriol. (Orig. B)*, 157, 34, 1973.

266. Schmiedel, G., Filser, J. G., and Bolt, H. M., Rat liver microsomal transformation of ethylene to oxirane in vitro, *Toxicol. Lett.*, 19, 293, 1983.

267. Filser, J. G. and Bolt, H. M., Exhalation of ethylene oxide by rats on exposure to ethylene, *Mutat. Res.*, 120, 57, 1983.

268. Bolt, H. M., Filser, J. G., and Störmer, F., Inhalation pharmacokinetics based on gas uptake studies. V. Comparative pharmacokinetics of ethylene and 1,3-butadiene in rats, *Arch. Toxicol.*, 55, 213, 1984.

269. Filser, J. G. and Bolt, H. M., Inhalation pharmacokinetics based on gas uptake studies. VI. Comparative evaluation of ethylene oxide and butadiene monoxide as exhaled reactive metabolites of ethylene and 1,3-butadiene in rats, *Arch. Toxicol.*, 55, 219, 1984.

270. Farmer, P. B., Bailey, E., and Shuker, D. E. G., The determination of in vivo alkylation of hemoglobin and DNA using gas chromatography-mass spectrometry, in *Developments in the Science and Practice of Toxicology*, Hayes, A. W., Schnell, R. C., and Miya, T. S., Eds., Elsevier, Amsterdam, 1983, 273.

271. Hamm, T. E., Guest, D., and Dent, J. G., Chronic toxicity and oncogenicity bioassay of inhaled ethylene in Fischer-344 rats, *Fundam. Appl. Toxicol.*, 4, 473, 1984.

272. Segerbäck, D., Alkylation of DNA and hemoglobin in the mouse following exposure to ethene and ethene oxide, *Chem. Biol. Interactions*, 45, 139, 1983.

273. Reid, D. M., Sheffer, M. G., Pierce, R. C., Bezdicek, D. F., Linzon, S. N., Reuvers, T., Spencer, M. S., and Vena, F., *Ethylene in the Environment: Scientific Criteria for Assessing its Effects on Environmental Quality*, National Research Council Canada, Ottawa, 1985.

274. Chandra, G. R. and Spencer, M., A micro apparatus for absorption of ethylene and its use in determination of ethylene in exhaled gases from human subjects, *Biochim. Biophys. Acta*, 69, 423, 1963.

275. Fuchs, Y. and Chalutz, E., Eds., *Ethylene: Biochemical, Physiological and Applied Aspects*, Martinus Nijhoff, Amsterdam, 1984.

276. Austin, S. G., Spontaneous abortions in hospital sterilising staff, *Br. Med. J.*, 286, 1976, 1983.

277. Gordon, J. E. and Meinhardt, T. J., Spontaneous abortions in hospital sterilising staff, *Br. Med. J.*, 286, 1976, 1983.

INDEX

A

U

V

W

Z

Printed and bound by CPI Group (UK) Ltd, Croydon, CR0 4YY

22/10/2024

01777630-0002